Understanding Chemistry

Understanding Chemistry

Chip Lovett

Raymond Chang

Williams College

 **Higher Education**

Boston   Burr Ridge, IL   Dubuque, IA   Madison, WI   New York   San Francisco   St. Louis
Bangkok   Bogotá   Caracas   Kuala Lumpur   Lisbon   London   Madrid   Mexico City
Milan   Montreal   New Delhi   Santiago   Seoul   Singapore   Sydney   Taipei   Toronto

Higher Education

UNDERSTANDING CHEMISTRY

Published by McGraw-Hill, a business unit of The McGraw-Hill Companies, Inc., 1221 Avenue of the Americas, New York, NY 10020. Copyright © 2005 by The McGraw-Hill Companies, Inc. All rights reserved. No part of this publication may be reproduced or distributed in any form or by any means, or stored in a database or retrieval system, without the prior written consent of The McGraw-Hill Companies, Inc., including, but not limited to, in any network or other electronic storage or transmission, or broadcast for distance learning.

Some ancillaries, including electronic and print components, may not be available to customers outside the United States.

 This book is printed on recycled, acid-free paper containing 10% postconsumer waste.

1 2 3 4 5 6 7 8 9 0 QPD/QPD 0 9 8 7 6 5 4 3

ISBN 0–07–255553–X

Publisher: *Kent A. Peterson*
Sponsoring editor: *Thomas D. Timp*
Senior developmental editor: *Shirley R. Oberbroeckling*
Senior marketing manager: *Tamara L. Good-Hodge*
Senior project manager: *Gloria G. Schiesl*
Production supervisor: *Kara Kudronowicz*
Senior media project manager: *Stacy A. Patch*
Senior media technology producer: *Jeffry Schmitt*
Senior coordinator of freelance design: *Michelle D. Whitaker*
Cover/interior designer: *Elise Lansdon*
Cover images: Blue flame and flask images: © *Photodisc, Volume 54*
Compositor: *The GTS Companies/Los Angeles, CA Campus*
Typeface: *10/12 Times Roman*
Printer: *Quebecor World Dubuque, IA*

Photo credit for page 10, "Dalton's chart of the elements": © Science & Society Picture Library

Library of Congress Cataloging-in-Publication Data

Lovett, Chip.
   Understanding chemistry / Chip Lovett, Raymond Chang. —1st ed.
      p.   cm.
   Includes index.
   ISBN 0–07–255553–X (acid-free paper)
      1. Chemistry. I. Chang, Raymond. II. Title.

QD33.2.L68   2005
540—dc22   2003066601
                                                            CIP

www.mhhe.com

In Memory of James F. Skinner (1940–1988)—

A Dedicated Chemical Educator and a Cherished
Friend

Chip Lovett
Raymond Chang

About the Authors

Chip Lovett

Chip Lovett was born in Long Island, New York, and began his college career studying architecture. During a seven-year hiatus from college he worked as a portrait artist, musician, truck driver, and postal worker. He returned to school and took his first chemistry course at the age of 26. He received his B.S. and M.S. in chemistry from California State Polytechnic University at Pomona, and his Ph.D. in biochemistry from Cornell University.

In 1985 he joined the chemistry department at Williams College, where he teaches courses on introductory chemistry, biochemistry, and AIDS. His major research interest is understanding how protein molecules control the way genes are turned on and off. His research at Williams has involved more than 70 undergraduate students and has resulted in numerous scientific publications dealing with the control of DNA repair genes.

He has served as Director of the Science Center at Williams since 1993 and directs summer science programs for children, elementary-school teachers, and incoming college students.

For relaxation, Professor Lovett plays the guitar and the piano and enjoys skiing, motorcycling, and horseback riding.

Raymond Chang

Raymond Chang was born in Hong Kong and grew up in Shanghai and Hong Kong, China. He received his B.Sc. degree in chemistry from London University, England, and his Ph.D. in chemistry from Yale University. After doing postdoctoral research at Washington University and teaching for a year at Hunter College of the City University of New York, he joined the chemistry department at Williams College, where he has taught since 1968.

Professor Chang has served on the American Chemical Society Examination Committee, the National Chemistry Olympiad Examination Committee, and the Graduate Record Examination (GRE) Committee. He is an editor of *The Chemical Educator*. Professor Chang has written books on physical chemistry, industrial chemistry, and physical science. He has also coauthored books on the Chinese language, children's picture books, and a novel for juvenile readers.

For relaxation, Professor Chang maintains a forest garden; plays tennis, Ping-Pong, and the harmonica; and practices the violin.

Contents

Preface

We wrote *Understanding Chemistry* specially for students taking chemistry for the first time, either in high school or college. It is not a textbook in the conventional sense because it does not have the encyclopedic coverage of topics that you will find in a general chemistry text. We have carefully selected areas that we believe are essential for building a sound foundation for learning and understanding chemistry. Because students are often intimidated by chemistry and by chemistry textbooks, we have used a number of strategies to make this book readable, user friendly, and pedagogically effective: familiar analogies to help explain abstract ideas; cartoons (created and drawn by Chip Lovett) to help students visualize chemical processes at the molecular level and to provide some comic relief in the learning process; and portraits of scientists describing their important discoveries to add a personal touch and to serve as a useful tool for remembering facts and concepts.

Because problem-solving is a major part in learning chemistry, we have provided a number of worked examples in the text. To learn how to solve problems you must practice, so we have created a website (www.understandingchemistry.com) with many worked examples and problems, as well as further discussion of topics. You will also find numerous colored molecular models and animations on the Web that will help you visualize the three-dimensionality of molecules and chemical reactions.

We hope you will find that this book is just what you need to get the most out of your introductory chemistry course, to help you understand concepts and do well on tests, and finally, to appreciate what a fascinating subject chemistry really is.

Features

- Unique cartoons lend themselves to understanding the chemistry behind the concept they are representing. They act as visual tools to help commit a chemical concept into memory.
- Analogies are used freely throughout the text to help understand abstract concepts.
- Worked examples take you through a step-by-step process to build problem-solving skills.
- A summarizing problem at the end of most chapters will tie topics together. The problem takes you through strategizing and solving an inclusive problem.

Media

An easy-to-use website has been created to complement *Understanding Chemistry* as a companion to the text.

- Practicing problems is a necessary requirement for succeeding in and mastering chemistry. A multitude of various problem types are provided on the site for student practice.
- Answers to the questions posted in the text and on the website are provided.
- Molecular models and animations to enhance the understanding of chemical concepts are presented.
- A periodic table with historical background is provided to understand the naming conventions of the elements.
- A database of links to other sites is provided for further information.

Acknowledgments

We would like to thank the following individuals, whose comments were of great help in preparing this first edition.

Aaron Brown
Los Angeles Community College

John Chrysochoos
University of Toledo

Fredesvinda Dura
LaGuardia Community College

Sharon Fetzer Gislason
University of Illinois

Wes Fritz
College of DuPage

Caroline Gill
Lexington Community College

John Gracki
Grand Valley State University

Tom Guetzloff
West Virginia State College

Ya-Ping Huang
Austin Community College

Silvia Kolchens
Pima Community College

Cliff LeMaster
Boise State University

Kristan Lenning
Lexington Community College

Gino A. Romeo, Jr.
Pima Community College

Pat Schroeder
Johnson County Community College

Thomas Selegue
Pima Community College

Mark Sinton
Clarke College

We thank our developmental editor Shirley Oberbroeckling who supervised the project, senior project manager Gloria Schiesl for taking care of the production details, and Jeff Schmitt and Stacy Patch for overseeing the website. Arthur Okwesili of Williams College provided much assistance in organizing the website. We also thank our editor Thomas Timp and publisher Kent Peterson for their encouragement and support in general.

Chip Lovett
Raymond Chang

A Student Walk-Through

nderstanding Chemistry is not a textbook in the conventional sense because it does not have the encyclopedic coverage of topics that you will find in a general chemistry text. Carefully selected areas are presented that are essential for building a sound foundation for learning and understanding chemistry.

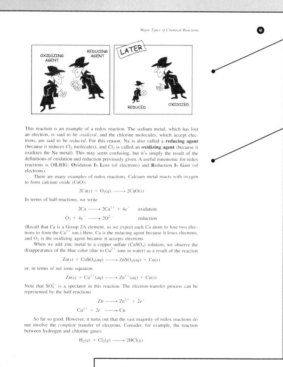

Unique Cartoons

Cartoons lend themselves to understanding the chemistry behind the concept they are representing. They act as visual tools to help commit a chemical concept into memory.

Analogies

Used freely throughout the text, analogies are another source of learning and retaining chemical knowledge.

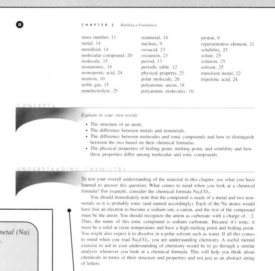

Worked Examples

Worked examples take you through a step-by-step process to guide you through similar problems.

Synopsis Problems

A synopsis problem at the end of most chapters tie the chapter together. The problem takes you through strategizing and solving an inclusive problem.

Media Walk-Through

n easy-to-use website has been created to complement *Understanding Chemistry* as a companion to the text.

Practice Problems

Practicing problems is a necessary requirement for succeeding and mastering chemistry. A multitude of various problem types are provided on the site for student practice.

Answers to the questions posted in the text and on the website are provided.

> **Example 3.16** Titanium is a strong, lightweight, corrosion-resistant metal that is used in rockets, aircraft, jet engines, and bicycle frames. It is prepared by the reaction of titanium(IV) chloride with molten magnesium between 950°C and 1150°C:
>
> $$TiCl_4(g) + 2Mg(l) \longrightarrow Ti(s) + 2MgCl_2(l)$$
>
> In a certain industrial operation 3.54×10^7 g of $TiCl_4$ are reacted with 1.13×10^7 g of Mg. (a) Calculate the theoretical yield of Ti in grams. (b) Calculate the percent yield if 7.91×10^6 g of Ti are actually obtained.
>
> **Reasoning and Solution** We follow the procedure in Example 3.15 to find out which of the two reactants is the limiting agent. This knowledge will enable us to calculate the theoretical yield. The percent yield can then be obtained by applying Equation (3.4).
>
> (a) First we calculate the number of moles of $TiCl_4$ and Mg initially present:
>
> $$\text{moles of } TiCl_4 = 3.54 \times 10^7 \text{ g } TiCl_4 \times \frac{1 \text{ mol } TiCl_4}{189.7 \text{ g } TiCl_4} = 1.87 \times 10^5 \text{ mol } TiCl_4$$
>
> $$\text{moles of Mg} = 1.13 \times 10^7 \text{ g Mg} \times \frac{1 \text{ mol Mg}}{24.31 \text{ g Mg}} = 4.65 \times 10^5 \text{ mol Mg}$$
>
> Next, we must determine which of the two substances is the limiting reagent. From the balanced equation we see that 1 mol $TiCl_4 \Leftrightarrow 2$ mol Mg; therefore, the number of moles of Mg needed to react with 1.87×10^5 moles of $TiCl_4$ is

Molecular Models/Animations

Molecular models and animations to enhance the understanding of chemical concepts are presented.

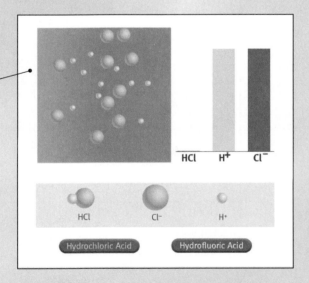

1

In the Beginning

Read this first!

In the beginning there were **atoms,**[1] **molecules,** and **ions.** After years of thinking about the world around them, people from every corner of Earth began to explore and discover the properties of the stuff we call matter. From these discoveries sprung chemistry—the science of matter. The first chemical experiments (performed thousands of years ago) probably involved testing the properties of different metals to make better weapons or testing the ability of plant extracts to treat illness. Priorities haven't changed much since then. With the eventual discovery of atoms and **compounds,** it seemed that it would be possible to actually understand why the material components of life and Earth behave the way they do. As scientists became intent on understanding the properties of substances, chemical experiments became more focused on testing hypotheses about the nature of matter. Scientists and students alike were excited about the subject of chemistry. Along the way, an enormous amount of knowledge about the structure and properties of atoms, molecules, and ions was acquired. To communicate this new-found knowledge, new words had to be invented to describe previously unknown substances and their behavior. Using this new chemical language, scientists began describing in textbooks the knowledge gleaned from countless experiments—and the textbooks kept growing in size. Textbooks for introductory chemistry are now so full of information that it would be practically impossible for anyone to learn it all in a year. This situation presents a challenge for chemistry teachers because students can't discern what information is essential to understanding chemistry and what information is not as important. To *understand* chemistry, a student needs to learn only a handful of concepts and principles that form the foundation for the science of matter—living and nonliving alike.

We designed this book for students taking chemistry for the first time, either in high school or college. It's also for students who have studied some chemistry but feel they really don't understand it. It's especially for students who find chemistry formidable, intimidating, scary, difficult, or just plain impenetrable. As chemistry teachers, we've encountered too many students who believe they really can't understand chemistry—students who take the introductory course like a bitter pill, hoping to come out of it with a decent grade. Of course, many students avoid the subject altogether. (This wasn't possible 100 years ago when all college students had to take chemistry!) We wrote this book because we honestly believe that everyone can understand chemistry at the introductory level and because we want to help students

---

1. Words in boldface are defined in the glossary and described in later chapters.

appreciate a subject that provides the basis for understanding everything that deals with matter.

The Three Objectives of This Book

This book was written with three objectives in mind: (1) to focus only on those topics that are *essential* to really understand chemistry, (2) to help the readers through the usual *hurdles* in a beginning chemistry course in ways that will make sense to them, and (3) to present the subject in a way that the readers *understand, remember, and enjoy.* We believe these are the most important objectives for helping everyone who uses this book to master all of the material that we describe, to do well in his or her chemistry course, and to be on solid ground for more advanced topics.

The Essentials

Chemistry is the study of atoms, molecules, and ions—that is, what they look like (or their structure) and how they behave. Understanding chemistry means having a working knowledge of the structure of these elementary particles and how (and why) they interact with one another. The good news is that both the structure and behavior of atoms, molecules, and ions are logical and in most cases predictable, and we will do our best to convince you of this fact. In this book, we have chosen to present only the information that we think is absolutely essential for you to acquire this knowledge.

The Hurdles

Chemistry deals with a world so tiny that it's invisible to the casual observer. Although chemists have devised equipment and techniques to observe the atoms and molecules that make up this tiny world, most people are still not comfortable thinking about things that they can't see. The first hurdle you must overcome in understanding chemistry is to learn to visualize these tiny particles—to think of them as objects with shapes and sizes (all small, of course). This is particularly important because it will help you deal with the assortment of abstract symbols, formulas, and equations—all couched in a foreign language. Without imagining what molecules look like, chemistry students can only associate unfamiliar words with an unfamiliar collection of symbols, formulas, and equations. Understanding chemistry on these terms would be extremely difficult and definitely no fun. You have to associate the language of chemistry and the symbols, formulas, and equations with the images of atoms and molecules themselves. Imagine trying to learn the word for "apple" in a foreign language if you have never seen (or eaten) an apple; the word would have little meaning because you wouldn't know what an apple is.

Another hurdle you must overcome is looking for and focusing on the common conceptual themes that relate all of chemistry. There is so much information in general chemistry textbooks that it's really hard for the beginning student to sort out what's essential for understanding the subject and what are merely details, trivia, and exceptional cases. Textbook authors are obliged to provide all of the information relevant to an introduction to chemistry (and we're glad they do), but you don't have to know it all to understand chemistry. Despite the best efforts of textbook authors and chemistry teachers, students often view all this information as an enormous collection of unrelated facts. In fact, there are only a few underlying themes that form the basis of chemistry, and once you understand them, the rest is relatively easy to add to the foundation of this knowledge. You may even find the details, trivia, and exceptional cases pretty interesting (as we do).

The third hurdle you must overcome is solving quantitative problems. Chemistry is an experimental science based in large part on drawing conclusions from measurements. Although general chemistry courses require some relatively straightforward calculations, students often see each problem as new and different. If this were the case, one would have to learn hundreds of different approaches to solve homework and exam problems. Again, chemistry would be an unreasonably difficult subject if this were the case. On the contrary, there are only a few different types of problems in general chemistry, but many variations of each type. So you need to learn only a few procedures (or approaches), but you also must learn to recognize the method needed to solve a particular problem.

Understanding, Remembering, and Enjoying Chemistry

The design of this book was inspired by our observation that students learn best when they enjoy the process and when they can engage themselves with the subject in a familiar and interesting context. We wanted it to be both an enjoyable read as well as a clear and friendly guide. To accomplish this goal, we have included pictures, some history, and many commonplace analogies to embellish the concepts.

Because chemistry deals mostly with things we can't normally see, the use of visual aids is especially important in understanding and remembering the subject. We have included many figures and cartoons to elaborate the formulas, equations, and text, and to add a human touch to the subject. These visual supplements will help you understand the concepts, and we hope they will also help you remember them. Another way to make chemistry memorable is to personalize it. Science is ultimately a very personal enterprise because scientists are people with their own individual pleasures, dreams, and biases. We have included information and pictures of some of the scientists who have pioneered the study of chemistry in the hope that this, too, will help you remember the concepts they helped develop.

There are many concepts in chemistry that are pretty abstract, and their development by scientists has often involved insights based on everyday experiences. Because it's always easier to understand something unfamiliar by relating it to something familiar, we have provided common analogies to illustrate the more abstract concepts. You may be surprised to learn that even the most brilliant scientists try to understand their work in the simplest terms. Many scientific discoveries might never have been made if scientists didn't try to reduce their results to the simplest possible explanation. The 14th-century scholastic monk William of Ockam has been cited by many generations of scientists for his insistence on economical explanations. He is often credited with saying, "If two theories explain the facts equally well, then the simpler theory is to be preferred." This principle, known as Ockam's razor, has inspired many scientists. In this book we have reduced introductory chemistry to the simplest explanation of the facts. We have also limited our focus on those facts that are essential to achieving a basic understanding of the structure and properties of atoms, molecules, and ions. Most of what you will learn is just variations on a few basic themes.

What Will You Have to Do to Understand Chemistry?

If you want to understand general chemistry (and do well on your exams), you will have to do the following:

- Learn to speak the language.
- Learn a handful of conceptual themes.
- Learn how to recognize and solve a few different types of quantitative problems.

The Language

Speaking the chemical language involves learning the names of chemicals and learning the terms used to describe fundamental processes and concepts.

Learning the names for chemicals turns out to be much easier than learning the names of your friends, acquaintances, and teachers. (Chemists use the term *nomenclature* for the naming system; it's from the Latin word *nomenclatura,* meaning "a list of names.") Learning the names of people is pure memorization—you have to memorize every individual's name because there are no rules that tell you a particular face should have a particular name (and it doesn't make it easier that some people have the same first name).

In chemistry there are simple rules for naming chemicals from a chemical formula (e.g., CO_2); without these rules, it would be impossible for anyone to learn the name of every chemical. You will have to memorize the rules, but there aren't very many. You will also have to memorize the names of a relatively small number of atoms and molecular species, and you may already know the names of some of these. The rules are given in Chapter 2, along with some useful tricks for memorizing the names of some molecular species. You can be sure that your first exam will have a question or two on nomenclature, and there is no good reason to lose any points on them, because it's not that hard.

You probably already know the names for some fundamental processes like melting, freezing, and boiling, as well as the names of some concepts like temperature, pressure, and heat. But you probably haven't thought about these concepts in terms of atoms and molecules (from which they ultimately derive their effects). Throughout this book, we will highlight all of the terms you should know and understand. These are words that can be found in the glossary of this book. Test yourself at the end of each chapter by trying to explain as much as you can about the highlighted words in the chapter.

The Conceptual Themes

Chemists are interested in understanding three fundamental properties of chemicals— their *structure,* their *physical behavior,* and their *chemical behavior.* Now that might sound rather dry, but to appreciate the relevance of what chemists do, you only have to realize that chemicals comprise everything—some chemists study water, some study DNA, some study drugs, and some study molecules that may someday run our computers. There are certain conceptual themes that underlie each of these three fundamental properties, and the structure and behavior of the millions of chemicals in the world are simply variations on these themes. Your mastery of general chemistry depends on learning these themes and how to recognize them in the chemical world. **An overarching theme is that for every substance the three fundamental properties are all interrelated.** A good exercise to practice throughout your study of chemistry is to ask how these properties are interrelated for specific substances. We will remind you of this point throughout the book.

Structure

What do molecules look like? The study of chemical structure in an introductory course mainly involves small molecules, but big molecules like antibiotics, proteins, and DNA are built with precisely the same structural features as for small molecules. Molecular structure is nothing more than simple geometry in which atoms are connected to each other by chemical bonds. There are essentially only five basic geometries that govern the construction of most molecules (and a majority of the molecules in the world use only three of them). The key to understanding structure is learning how atoms are connected to form a particular geometry. In general, we will be concerned with only a handful of different atoms. Moreover, this process is further simplified when you realize that atoms in the same vertical column in the **periodic table** usually form similar geometric structures when they combine with other atoms.

Physical Behavior

The physical behavior of chemicals includes everything that happens to a substance without changing its identity. As you will see later in this book, these are properties that don't involve breaking or making chemical bonds. For example, melting ice is physical behavior (or a **physical property**) because you start with water molecules in solid form and produce identical water molecules in the liquid state. Another physical property of water is its ability to dissolve sugar. A more elaborate example of a physical property is the interaction of a protein hormone with a site on DNA to turn on the expression of a particular gene. All of these examples, and many more, are based on one conceptual theme: **a positive charge attracts a negative charge**—like the opposite poles on a magnet. This electrostatic attraction underlies the forces that govern the physical behavior of matter, as well as the formation of the thousands of different ionic compounds.

Chemical Behavior

The chemical behavior (or **chemical property**) of substances is usually described in terms of **chemical reactions,** or what happens when one or more substances interact to produce other substances. In a chemical reaction, the bonds that hold atoms together in molecules are broken, and new bonds are formed. If you think of chemical reactions as processes that are accompanied by flames or explosions, you are already aware of a conceptual theme that is associated with every chemical reaction: **Energy is either produced or consumed in chemical reactions.**

To understand chemical reactions, it is especially important for you to realize that **most chemical reactions involve either the transfer of protons or the transfer of electrons**—those elementary particles that you may already be familiar with as the components of all atoms. (Chapters 2 and 6 describe these particles in the context of atomic structure.) Thus, we have two types of processes, proton transfer and electron transfer, that form the basis for nearly all chemical reactions.

An important concept inherent to both physical and chemical processes is **chemical equilibrium.** It applies to every chemical reaction that is reversible (and most of them are). A simple example of equilibrium is a closed container filled partially with water. Because the water molecules are constantly moving around, some at the surface will have enough energy to break away from the forces holding them in the bulk liquid. These free molecules become part of water vapor above the liquid, but the freed molecules can collide with the water and become part of the liquid again. Eventually, the number of water molecules in the vapor is sufficient to cause the rate of molecules returning to the water to equal the rate of molecules leaving the water. At this point, the number of water vapor molecules remains constant, and the system is at equilibrium. The situation is similar for a chemical reaction in which two molecules (the reactants) collide with one another to form two different molecules (the products). If the reaction is reversible, products can also collide with one another to form the same reactants. The difference between the evaporation and condensation of water and the chemical reaction is that the latter involves changing the identity of the molecule by breaking and making chemical bonds, while the former involves only breaking the forces that hold molecules together in the liquid or solid state. Every equilibrium situation involves one or the other or both—just variations on the same themes.

Thus, the conceptual themes that are important in general chemistry include the following:

1. Structure, physical properties, and chemical properties are interrelated.
2. Structure is simply geometry in which atoms are connected by chemical bonds.
3. A positive charge attracts a negative charge.
4. Energy is either produced or consumed in chemical reactions.
5. Most chemical reactions involve either the transfer of protons or the transfer of electrons.
6. Many chemical reactions reach equilibrium where the amounts of reactants and products remain constant.

The Problems

Chemistry is an experimental science that relies heavily on taking measurements at the macroscopic level to determine what's happening at the molecular level. Most of the problems you will have to solve in general chemistry involve converting macroscopic measurements into information about molecular details. The good news is that you will encounter only a few different types of problems, and we will show you how to systematically solve them using straightforward approaches. You will then need to learn to recognize only the particular type of problem.

There are three keys to problem solving: practice, practice, practice. You should keep at it until you are confident that you can solve any problem of each type. Your chemistry teacher will assign plenty of problems, and if you need more practice, we have provided lots of them on the website. To master problem solving in chemistry, you must realize and constantly remind yourself that these are all variations on a relatively small number of different problem types.

Students have trouble with chemistry problems because they either don't practice and/or they treat every problem they encounter as something requiring a new and totally different solution strategy.

Getting the Most Out of Your Chemistry Class

We assume that you are reading this book because you are taking a chemistry class. Although the expectations and testing methods may differ somewhat among chemistry classes, the following suggestions should maximize both your understanding of the course material and your performance on exams in any course.

Attend class. Get there for the start of the class when important announcements are made or handouts are distributed. It's also a good idea to sit in the front row. You will be less distracted by others in the class, and if for some reason your final grade ends up on the borderline between an A and a B, your teacher is more likely to look at your case favorably.

Take careful notes and be sure you understand them. Your understanding of the material introduced in class will be greatest if you take the time to read through your class notes on the same day of class, when the lecture (including any demonstrations) is still fresh in your mind. At that time you should note points of confusion and get them cleared up (preferably, with your teacher) before the next class period. **The best way to be sure that you understand the material is to try to explain it to someone else.** The old adage that you never really learn something until you teach it is often true with chemistry. Not only will explaining the material in your own words reinforce your understanding of it, you will easily identify the parts you don't understand.

Read assigned sections from the textbook before the class in which the topics will be discussed. Make notes in the margins of your text, especially for topics that you don't completely understand. Why do you think they put those big margins in chemistry textbooks? Throw away your highlighters and write extensively in your textbook. Making your own notes is a much more active process and will prime you for getting the most out of the lecture. You might find it helpful to refer to the text again after class to help clear up points discussed in lecture.

Do the assigned homework problems systematically as the material is covered in class. Promptly work on problems that relate to material just covered in a lecture. You should also refer to the relevant examples in your text when working on problems. It's often helpful to work on problems with your classmates, provided everyone is actively engaged in the process. This practice also gives you an opportunity to try your hand at explaining the material to someone else.

Don't confuse quantity of studying with quality of studying. Going into solitary confinement with your chemistry book for six hours or "pulling an all-nighter" is an extremely poor way to study (and a waste of time). Most people can't concentrate effectively on the same subject for more than two to three hours at one stretch. Studying chemistry for two hours on three successive days will likely result in a much better retention than putting in six hours on a single night (especially if the exam is the next day).

In reviewing homework problems prior to an exam, don't simply read through your answers to the problems. This method is very passive and teaches you little. Read the problem, cover your answer, and think logically through the steps you follow to answer the problem. In some cases, drawing a simple picture can help you organize your thoughts and better understand what is asked of you.

Review sessions with your classmates can be very helpful. While it is probably best to do much of your studying alone in a quiet place, review sessions with several students preparing for a quiz or an exam can be very effective. Students are usually more willing to expose their confusion to one another than to a teacher. And as we've said before, explaining a given point to another student is an excellent way to test and reinforce your own understanding of the subject.

Take advantage of your teacher's office hours. Your teacher is an extremely valuable resource. If you don't understand something, ask your teacher as soon as possible. The parts you don't understand will likely provide background for later material so don't wait to get it cleared up. Again, if for some reason your final grade ends up on the borderline between an A and a B, your teacher will be more likely to push it upward if you show up at office hours and demonstrate that you are serious about learning the material. Here's a useful hint: Before you appear in your teacher's office, make a list of the things that are unclear to you so you don't waste his or her time by flipping through pages of your textbook or notes, frantically trying to locate places that gave you trouble. Remember that the first step toward understanding a concept is being able to say what it is about the concept that you don't understand. It's also a good idea to get to know your teacher because someday you may want him or her to write a letter of recommendation on your behalf.

Your Teacher Rules

Two guys were taking chemistry at Williams College. They did well on all of the quizzes, midterms, and labs, and had a solid "A" going into the final. They were so confident that the weekend before finals (the chemistry final was on Monday), they decided to go over to Amherst College and party with some friends. They had a great time; however, they overslept on Sunday and didn't make it back to Williams until early Monday morning. Rather than take the final then, they found their professor after the final and explained to him why they missed it. They told him that they had gone to Amherst for the weekend and had planned to return in time to study, but they had a flat tire on the way back. They didn't have a spare and couldn't get help for a long time, and that's why they were late in getting back to campus. The professor thought this over and told them they could make up the final the following day. The two guys were elated and relieved. They studied that night and went in the next day to take the final. The professor placed them in separate rooms, handed each one a test booklet, and told them to begin. They each looked at the first problem, which was worth 5 points. It was a simple question about solution concentrations. "Cool," they thought. "This is going to be easy." They answered the question and then turned the page. They were not prepared, however, for what they saw on this page. It said: (95 points)—Which tire?

Building a Foundation
Learning to think like a chemist

How many times can you divide a piece of gold wire until it isn't gold anymore? The 5th-century B.C. Greek philosopher *Democritus* (who, according to Aristotle, "thought about everything") proposed that you couldn't do this forever because all matter consists of very small, indivisible particles. He called these particles *atomos,* meaning "uncuttable" or "indivisible." It was not until 1808 that an English scientist and school teacher, *John Dalton,* formulated a precise definition of the indivisible building blocks of matter that we call **atoms.** While Democritus's idea of the atom was purely theoretical, Dalton's formulation was based on his and other people's experimental work on atomic masses (also called atomic weights) and their relationships in substances that contain different types of atoms. This combination of experiment and theory marked the beginning of the modern era of chemistry, elevating the discipline to a quantitative science. Dalton's atomic theory stimulated the rapid progress of chemistry during the 19th century and forms the basis of much of the chemistry we do today. The theory can be summarized as follows:

- **Elements** are made of identical, tiny particles called atoms, and the atoms of one element are different from the atoms of all other elements.
- **Compounds** are made of atoms of more than one element combined in whole number ratios.
- In **chemical reactions,** atoms can only separate, combine, or rearrange; they can't be created or destroyed.

Atoms

Atoms are really small, and atoms of one element are different from atoms of another element. But they are not indivisible (otherwise, chemistry as a discipline would not exist). Experimental evidence gathered over many years (mostly by physicists) has established the following facts about an atom. It's shaped like a sphere but is mostly empty inside. At the center of the sphere is a space called the **nucleus** that contains two types of particles—protons and neutrons. The **proton** is positively charged

> MATTER IS MADE OF TINY PARTICLES THAT CANNOT BE DIVIDED. I CALL THEM ATOMOS, WHICH IS GREEK FOR UNCUTTABLE.

Democritus believed in atoms a long time ago.

Dalton's atomic theory
may have been inspired by his
favorite sport—lawn bowling.

Dalton's chart
of the elements.

and is assigned a charge of $+1$ while the **neutron** has no charge. These two nuclear particles are about the same mass, with the neutron being a little heavier. The space occupied by protons and neutrons (that is, the nucleus) is extremely small compared to the total volume of the atom.

A third type of atomic particle is the negatively charged **electron,** which has a charge of -1. The electrons are spread out around the nucleus and at some distance from it. For now, the two things to remember about atoms are

1. **Because atoms are electrically neutral, the number of protons is equal to the number of electrons.**
2. **All chemical reactions involve either the loss and gain of electrons or the sharing of electrons between atoms.** (You see, atoms are not really indivisible.)

The mass of an electron is about 1840 times smaller than the mass of a proton. In fact, a neutron can be split to produce a proton and an electron (but that's another story).

The discovery that atoms are comprised of a nucleus containing protons and neutrons surrounded by electrons and a lot of empty space was made in 1910 by the physicist **Ernest Rutherford.** (Although Rutherford claimed that all science was either physics or stamp collecting, he actually won the Nobel Prize in CHEMISTRY!) This critical discovery was made when he and his collaborator **Hans Geiger** (of Geiger counter fame) and an undergraduate student named **Ernst Marsden** fired positively charged alpha particles (helium atoms stripped of their electrons) into gold foil and measured their deflection. They found that most particles went through with little or no deflection. But once in a long while, an alpha particle would bounce back in the direction from which it came; this really surprised Rutherford who said, "It was as incredible as if you had fired a 15-inch shell at a piece of paper and it came back at you."

Rutherford discovers the nucleus.

Rutherford reasoned the positively charged alpha particles were deflected by positive charges within the atom. The fact that most of the particles were undeflected suggested the atom was largely empty space (sort of like firing rockets through our solar system where only the ones that hit the sun would be deflected). Rutherford therefore proposed that all of the atom's positive charges (that is, the protons) were concentrated in a very small space, which he called the nucleus.

We now know that the nucleus, where 99.99% of the mass is located, occupies only a tiny fraction of the atom's space. You can appreciate the relative sizes of an atom and its nucleus by imagining that if an atom were the size of the Houston Astrodome, the volume of the nucleus would be comparable to that of a small marble. This means that the volume of atoms is about 99.99% empty space (and because we are made entirely of atoms, we also are about 99.99% nothing!) and that the nucleus is really, really dense. How dense is it? Imagine squeezing about 200 million African bull elephants, tusks and all, into the space occupied by a small marble and you'd be pretty close to the density (mass per unit volume) of the atomic nucleus.

The number of protons in an atom (which we call the **atomic number**) defines an element because the atoms of a particular element have a specific number of protons. For example, the atoms of hydrogen have one proton or an atomic number of 1, carbon has six protons and an atomic number of 6, and uranium has 92 protons and an atomic number of 92. The **mass number** of an atom is the combined number of protons and neutrons present. Chemists use a simple shorthand notation to represent elements where the mass number (number of protons and neutrons) is a superscript to the left of the chemical symbol and the atomic number (number of protons) is a subscript:

$$_1^1\text{H} \qquad _6^{12}\text{C} \qquad _{92}^{238}\text{U}$$

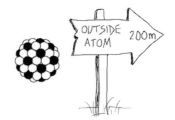

If the nucleus were the size of a marble, the radius of the atom would be 200 meters.

Why do we do this? Do we really need to identify an atom by both its number of protons and neutrons? As it turns out, atoms of the same element usually come in different forms where the differences are due to different numbers of neutrons. Atoms with the same atomic number but different numbers of neutrons are called **isotopes,** and some of these are radioactive because they're not very stable (but that again is another story). The number of neutrons present doesn't affect the identity of an element or its chemical properties, but it does change the mass of an atom. What is the role of neutrons? Because the protons are packed very tightly in a nucleus, you can imagine there would be considerable repulsion between like (positive) charges. The neutrons are there to hold things together and keep the nucleus intact. Hydrogen has three isotopes, which contain one proton and no neutron, one proton and one neutron, and one proton and two neutrons (Figure 2.1). So if we know the mass number of an atom, we will know the number of neutrons present

Figure 2.1 Nuclei of the three isotopes of hydrogen. Each isotope would have one electron at some distance away.

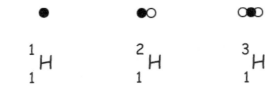

$${}^{1}_{1}H \qquad {}^{2}_{1}H \qquad {}^{3}_{1}H$$

by subtracting the atomic number from the mass number. How do we know the number of electrons? (Remember that the number of electrons is equal to the number of protons in an atom.)

2.1

EXAMPLE

Radon gas has been in the news much these days because of its detrimental effects on human health. How many neutrons are present in the radioactive isotope ${}^{222}_{86}Rn$?

Answer   *The larger number (222) must be the mass number, and the atomic number is 86. The number of neutrons is therefore given by (222 − 86) = 136.*

2.2

EXAMPLE

A student wrote two incomplete symbols for an isotope of magnesium: ${}^{25}Mg$ and ${}_{12}Mg$. Which symbol is the more informative? Why?

Answer   *Because we are given the name of the element, we can use the table of names of elements given in the inside cover to determine the atomic number of magnesium, which is 12. But normally we would have no idea about the number of neutrons present. Therefore, ${}^{25}Mg$ is more informative because it tells us that the number of neutrons is (25 − 12) = 13.*

Check out some of the sample problems on the website if you'd like some practice.

At the time we are writing this book, there are 113 distinct elements known—there may be more when you read it. That may seem like a lot, but chemistry gets much simpler when you focus on the common ones; about 95% of the compounds on this planet contain only a small fraction of the known elements. But even if you had to learn something about all of them (which fortunately you don't), it is still easier than it might seem when you realize that a lot of elements behave alike. In fact, the biggest bonus in the study of chemistry is that the chemical properties of elements are periodic functions of their atomic numbers. What does that mean? Well, if you were to line up the elements in order of increasing atomic number, you'd find that the 2nd, 10th, 18th, 36th, 54th, 86th all have similar chemical behavior. This is the basis for the **periodic table,** the single most important tool in general chemistry.

The Periodic Table

The periodic table is the ultimate guidebook for students of chemistry. The elements are arranged by atomic number (the number above the element symbol) in horizontal rows called **periods** and in vertical columns known as **groups** or **families,** according to similarities in their chemical properties. Note that elements 110–112 and 114 have recently been synthesized, although not all of them have been named. For a while we thought we had elements 116 and 118 too, but it turned out to be falsely claimed. Sometimes that's the way it goes in science.

Take a look at the periodic table on the inside cover of your book. The present form of the periodic table was proposed by the Russian chemist ***Dimitri Mendeleev*** in 1869,

Mendeleev arranged cards for each of the known elements in groups on a table as if he were playing his favorite game of solitaire.

	Gruppe I.	Gruppe II.	Gruppe III.	Gruppe IV.	Gruppe V.	Gruppe VI.	Gruppe VII.	Gruppe VIII.
Typische Elemente	H 1 Li 7	Be 9,4	Bo 11	C 12	N 14	O 16	F 19	
Reihe 1	Na 23	Mg 24	Al 27,3	Si 28	P 31	S 32	Cl 35,5	
- 2	Ka 39	Ca 40	—44	Ti 50(?)	V 51	Cr 52	Mn 55	Fe 56, Co 59, Ni 56, Cu [63
Reihe 3	(Cu 63)	Zn 65	—68	—72	As 75	Se 78	Br 80	
- 4	Rb 85	Sr 87	(Yt 88)(?)	Zr 90	Nb 94	Mo 96	—100	Ru 104, Rh 104, Pl 106, [Ag 108
Reihe 5	(Ag 108)	Cd 112	In 113	Sn 118	Sb 122	Te 125	J 127	
- 6	Cs 133	Ba 137	—137	Ce 138 (?)	—	—	—	
Reihe 7	—	—	—	—	—	—	—	
- 8	—	—	—	—	Ta 183	W 184	—	Os 199 (?), Jr 198, Pt [197, Au 197
Reihe 9	(Au 197)	Hg 200	Tl 204	Pb 207	Bi 208	—	—	
- 10	—	—	—	Th 232	—	Ur 240	—	
Höchste salz- bild. Oxyde	R^2O	R^2O^2 od. RO	R^2O^3	R^2O^4 o. RO^2	R^2O^5	R^2O^6 o. RO^3	R^2O^7	R^2O^8 od. RO^4
Höchste H- Verbindung				RH^4	RH^3	RH^2	RH	(R^2H) (?)

1. Periode 2. Periode 3. Periode 4. Periode 5. Periode

Mendeleev's periodic table

although several other chemists also played around with the same idea. Mendeleev saw that if the elements known in his time were arranged in order of increasing atomic mass, certain properties (like corrosive metals that react violently with water) recurred at regular intervals. But he found holes in his table because some elements had not yet been discovered. In fact, his periodic table was not widely accepted by chemists until the missing elements—gallium, germanium, and scandium—were discovered and shown to have the properties that Mendeleev predicted. (When Mendeleev was buried in St. Petersburg in 1907, his students carried a chart showing the periodic table in his funeral procession.) Discoveries made in the first half of the 20th century regarding the arrangement of electrons around the nucleus make it easy to understand why the elements would show this type of chemical periodicity. We will discuss this fact in Chapter 6.

All the elements can be divided into three categories—**metals, nonmetals,** and **metalloids,** respectively, on your periodic table. A metal is a good conductor of heat and electricity because the outermost electrons are mobile; a nonmetal is usually a poor conductor of heat and electricity because the electrons in them are not free to move around.

A related property that distinguishes metals from nonmetals is their tendency to lose or gain electrons. Metals have a weak hold on their outermost electrons and will lose them easily when they react. By contrast, nonmetals hold on to their outermost electrons very tightly, and most of them will grab electrons from metals if given a chance.

A metalloid has properties that are intermediate between those of metals and nonmetals. The majority of known elements are metals; only 17 elements are nonmetals, and eight elements are metalloids. From left to right across any period, the physical and chemical properties of the elements change gradually from metallic to nonmetallic, so a region of elements that are intermediate in properties is exactly what you'd expect. (When you understand chemistry, you'll see that a lot of it is just what you'd expect.) Periodic tables will often mark a border between those elements that behave primarily as metals and those that behave primarily as nonmetals (the dark zigzag line on your periodic table). You will find this distinction useful later in Chapter 3 where we predict the formation of molecular and ionic compounds. The notable exception is hydrogen, which is not a metal even though it's on the left. Hydrogen typically will gain electrons from metals and lose or share them with nonmetals. *(Check out the website for some sample problems on classification of elements.)*

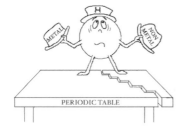

Elements are often referred to collectively by their periodic table group number (Group 1A, Group 2A, and so on). However, for historical reasons, some element groups are still known by special names. The Group 1A elements (Li, Na, K, Rb, Cs, and Fr) are called the **alkali metals,** and the Group 2A elements (Be, Mg, Ca, Sr, Ba, and Ra) are called the **alkaline earth metals.** Nonmetallic elements in Group 7A (F, Cl, Br, I, and At) are known as the **halogens,** and those in Group 8A (He, Ne, Ar, Kr, Xe, and Rn) are called the **noble gases.** *(Check out the website for the historical reasons for these names.)*

The periodic table is a terrific tool because it correlates the properties of the elements in a systematic way and helps us to make predictions about chemical behavior. We will frequently refer to it in this book, and it will be invaluable for homework problems and exams in your class. The table also greatly simplifies the understanding of chemistry because in most cases the elements in the same group have similar chemical properties. As we said earlier, this is a big bonus in studying chemistry. Most of the chemistry you will learn in general chemistry involves only a small number of elements. You will see a number of 1A and 2A metals and aluminum, which is the first metal in Group 3A. Because the elements in the same group have similar chemical behavior, you simply need to get a solid understanding of the chemistry (that is, how the atoms react) of H, He, Na, Mg, Al, C, N, O, F—the rest are mostly variations of each group.

Atoms React to Form Molecules and Ions

Atoms react with one another to form molecules or ions, and these chemical reactions and the nature of their products are the basis of chemistry.

Atoms often like to react with one another to become more stable. We like to be stable, but what does it mean for an atom to be stable? For now, you can think of stability (which can also be thought of as low potential energy) as the opposite of reactivity. Stable atoms are not reactive, and reactive atoms are not stable. Because the noble gases (Group 8A) exist as isolated atoms and are not typically combined with other atoms, we infer that they do not want to react. Therefore we say they are stable. Other atoms do not exist stably on their own and they react by losing or gaining electrons or by sharing electrons. Remember that **all chemical reactions involve either the loss and gain of electrons or the sharing of electrons between atoms.** What is it about the noble gases that makes them so stable? If we knew this, we could understand why and how atoms react. As we will discuss later (in Chapter 6), all of the noble gases (except helium) have eight electrons in their outer layer, or shell, as chemists usually call it. All other atoms (except hydrogen) tend to react to achieve a total of eight electrons in their outermost shell. Hydrogen often reacts to get a total of two electrons, like helium; however, sometimes hydrogen also loses its only electron.

For now, let's just focus on the products of chemical reactions: molecules and ions, and we'll return to chemical reactions in Chapter 3.

Molecules

A **molecule** is a combination of at least two atoms in a specific spatial arrangement held together by attractive forces called *chemical bonds*. This definition is a mouthful, but it is really important that you understand exactly what it means: a molecule is simply a bunch of tiny spheres (that is, atoms) stuck together with chemical glue. Molecules form when atoms react by sharing electrons. The shared electrons are the chemicals bonds (that is, the glue), and each bond is made up of two electrons. We

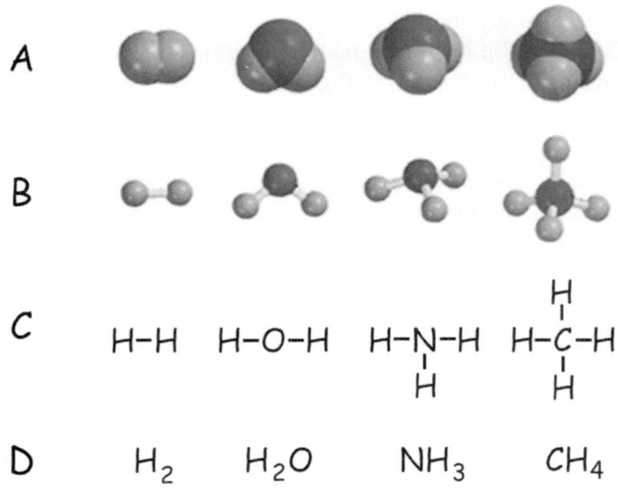

Figure 2.2   Hydrogen, water, ammonia, and methane molecules.

(a) space-filling models: what they really look like
(b) ball (atom) and stick (bond) models
(c) structural formulas
(d) molecular formulas

A

B

C   H–H   H–O–H   H–N–H   H–C–H

D H_2 H_2O NH_3 CH_4

want you to always imagine what molecules look like because not doing so means you won't be able to truly understand chemistry. Figure 2.2 shows pictures of some simple molecules represented in different ways. The space-filling structure is the most accurate representation; however, the others are useful for understanding molecular geometry.

A molecule may contain atoms of the same element or atoms of two or more elements joined in a fixed ratio. If the atoms belong to different elements, then the molecule is also known as a **compound.** (Remember Dalton's atomic theory?) Hydrogen gas, for example, is an element, not a compound, because it consists of molecules made up of only H atoms. Water, on the other hand, is a molecule that contains hydrogen and oxygen in a ratio of two H atoms to one O atom. Therefore, it is a compound—a molecular compound. Like atoms, molecules are electrically neutral.

To save time when talking and writing about chemistry, we use **chemical formulas** to represent molecules and compounds. For example, instead of saying "The hydrogen molecule contains two hydrogen atoms," we simply write H_2, where the subscript represents the number of atoms present. To additionally provide information about the physical state of a chemical, we can also designate whether it's solid, liquid, gas, or dissolved in water (aqueous) using the letters (*s*), (*l*), (*g*), and (*aq*), respectively.

The vast majority of molecules contain more than two atoms. They can be atoms of the same element, as in ozone (O_3), which is made up of three atoms of oxygen. Or they can be combinations of two or more different elements. Molecules containing only two atoms are called **diatomic molecules,** and those containing more than two atoms are called **polyatomic molecules.** Like ozone, water (H_2O), ammonia (NH_3), and methane (CH_4) are polyatomic molecules. Note that when the number of a particular type of atoms present is one, the subscript 1 is not shown as, for example, the O in H_2O, the N in NH_3, and the C in CH_4. Chemical formulas are great tools for describing chemistry, but to understand the chemistry, you have to think about what molecules look like. Look again at the pictures of hydrogen, water, ammonia, and methane in Figure 2.2.

Which types of atoms are most likely to form molecules? For our purpose, molecules contain mainly H, C, N, and O atoms. One especially useful piece of information to know at this stage is the way these atoms are linked together in a molecule. **Of these common elements, remember that hydrogen forms only one bond, oxygen forms two bonds, nitrogen forms three bonds, and carbon forms four bonds** (as shown in Figure 2.2). Remember that these atoms react to be like a noble gas, and they will share electrons in bonds to do so. Thus, hydrogen with one electron fewer than helium will form one bond, and oxygen with two electrons fewer than neon will form two bonds. You can always refer to the periodic table to see how many electrons an atom would have to gain in order to have the same number of electrons as a noble gas; this number of electrons gives you the number of bonds H, C, N, and O atoms will form. In Chapter 7 we'll see why this is so, but knowing it now will be very helpful.

Take a look at your periodic table again and guess how many bonds sulfur, phosphorus, and silicon might want to form. Because bond formation is a chemical property, you would predict that sulfur would bond like oxygen, phosphorus like nitrogen, and silicon like carbon (because atoms in the same periodic table group have similar chemical properties). This is in fact the case, although phosphorus and sulfur can sometimes form more bonds (but these variations will be easier to understand later). As we have said, all chemical reactions involve either the loss and gain of electrons or the sharing of electrons between atoms. Molecules form when atoms share electrons through the formation of chemical bonds. What happens when an atom loses or gains electrons?

NOW THAT I'M SHARING ELECTRONS WITH THESE TWO HYDROGENS, I FEEL LIKE I HAVE 8 ELECTRONS IN MY OUTER SHELL.

Oxygen is stable
when it forms bonds
with 2 hydrogens to produce
a water molecule.

Ions

An **ion** is an atom or group of atoms that has a net positive or negative charge (because they have either lost or gained electrons). Atoms that lose electrons are called **cations** (they're positive), and those that gain electrons are called **anions** (they're negative).

Cations and anions are attracted to each other and held together by electrostatic forces (that is, forces between positive and negative charges) to form **ionic compounds.** The force of attraction between oppositely charged ions is similar to the force of attraction between opposite poles of a magnet. As we said earlier, metals want to lose electrons and nonmetals want to grab electrons. In most cases, cations are derived from metals and anions are derived from nonmetals. This is yet another bonus in studying chemistry, because the zigzag line on your periodic table tells you which atoms are metals and which are nonmetals. The ones to the left of the zigzag line form cations, and the ones to the right form anions (usually, but we don't have to dwell on exceptions to understand chemistry). Listed in the following are the

Positively charged cat-ion.

TWO ATOMS STUMBLE OUT OF A BAR AND ONE BUMPS INTO A TELEPHONE POLE AND YELLS, "OUCH! I LOST AN ELECTRON!" THE OTHER ATOM ASKS, "ARE YOU SURE?" AND THE BRUISED ATOM SAYS, "I'M POSITIVE."

HA! HA! HA! HA!

Nonmetals will grab electrons from metals if they have a chance.

After electron transfer, anion and cation are attracted to one another.

charges on some common **monatomic** cations and anions, that is, ions that contain only one atom:

Group	1A	2A	3A		5A	6A	7A
	Li^+				N^{3-}	O^{2-}	F^-
	Na^+	Mg^{2+}	Al^{3+}			S^{2-}	Cl^-
	K^+	Ca^{2+}					Br^-
		Ba^{2+}					I^-

Note that all the ions listed come from atoms that lose or gain electrons to have the same number as a noble gas. After losing electrons, the resulting cations have the same number of electrons as the preceding noble gas (e.g., Na^+ is like Ne). Similarly, after gaining electrons, the resulting anions have the same number of electrons as the following noble gas (e.g., Cl^- is like Ar). An easy way to remember the charges on these metal cations is to note that in each case the charge is the same as the group number. The charges on the anions are given by the group number minus 8. For example, the charge on the anion of chlorine is $7 - 8 = -1$. The common **polyatomic anions,** or anions containing more than one atom, are CO_3^{2-} (carbonate), OH^- (hydroxide), NO_3^- (nitrate), PO_4^{3-} (phosphate), and SO_4^{2-} (sulfate). We will refer to these five ions again in this chapter and introduce you to several others as well, so it would be helpful to start making a collection of flash cards right now and practice saying their names. The only important polyatomic cation you need to know in general chemistry is NH_4^+, which is called the ammonium ion. (Add this to your collection of flash cards.) Note that this cation does not contain a metal atom.

Ionic Compounds

Atoms don't usually lose or gains electrons unless they interact with other atoms. Let's imagine adding chlorine gas, Cl_2, to a piece of sodium metal, Na. The yellowish gas is blown through a tube into an enclosed glass vessel containing a chunk of dull gray metal, and there is an instant and violent reaction. After the explosive fireworks subside, the vessel contains nothing but a pile of white powder. The white solid is the product of this chemical reaction; what do you think it is? It's the ionic compound sodium chloride (NaCl), also known as table salt. How do we know the formula is NaCl and not $NaCl_2$ or something else?

The key to writing chemical formulas of ionic compounds is electrical neutrality. Because Na is a Group 1A element, which produces +1 ions (remember?), and chlorine is a group 7A element, which produces -1 ions ($7 - 8 = -1$), the compound that results is the simplest electrically neutral combination, NaCl. If the charges on the cation and anion are the same, the formula is easy to write as in the case of NaCl (sodium chloride), $CaSO_4$ (calcium sulfate), and $AlPO_4$ (aluminum phosphate). If the charges on the ions are not equal, then we need to do some simple algebra. For example, in magnesium hydroxide (commonly known as milk of magnesia), we need two hydroxide ions (OH^-); that is, two negative charges to balance the two positive charges on a single magnesium ion (Mg is a Group 2A element). So the formula is $Mg(OH)_2$. We will say more about naming ionic compounds a little later.

The way to recognize ionic compounds is to realize that metals like to lose electrons and nonmetals like to gain electrons. For example, the alkali metals, most of the alkaline earth metals, and aluminum readily lose electrons while nitrogen, oxygen, sulfur, and the halogens readily accept electrons. Thus, a compound containing one of these metallic elements and nonmetallic elements is very likely to be an ionic compound. In most cases (the notable exception being ammonium compounds), if a compound contains both metals and nonmetals, it's ionic. Otherwise, the compound is classified as a molecular compound because no ions are present.

2.3

EXAMPLE

Classify the following compounds as ionic or molecular: 1. SO_2, 2. LiF, 3. NH_4Cl, 4. CaI_2.

Answer

1. *Both S and O are nonmetallic elements; therefore, this is a molecular compound.*
2. *Li is an alkali metal, and F is a halogen, so LiF is an ionic compound.*
3. *N, H, and Cl are all nonmetals, but as we said earlier, the ammonium ion is NH_4^+. This makes NH_4Cl an ionic compound.*
4. *Ca is an alkaline earth metal and I is a halogen, so the compound is ionic.*

Check out the website for more problems.

The Names of Chemical Compounds

Naming chemical compounds is not as tedious and painful as it may seem if you learn a few simple rules.

One of the things you have to learn in studying chemistry is what to call compounds—you have to know the language. Formally this topic is called **chemical nomenclature,** and while it may not be essential to understanding chemistry, you will need the language to talk and write about what you know. There are now some 14 million known compounds, so there is no way that anyone can memorize all the names. And there is no need to do so either because there are systematic procedures that can help us name all kinds of compounds. We'll try to make it as painless as possible. Although you will learn the names of the compounds as we go along, it's useful to be familiar with a few simple rules.

For starters you must learn the names of some elements and their corresponding symbols. We recommend that you memorize those in Figure 2.3. You should check with your teacher to see if you need to memorize more of them. For all exams in general chemistry, you'll probably be given a periodic table (or there may be one

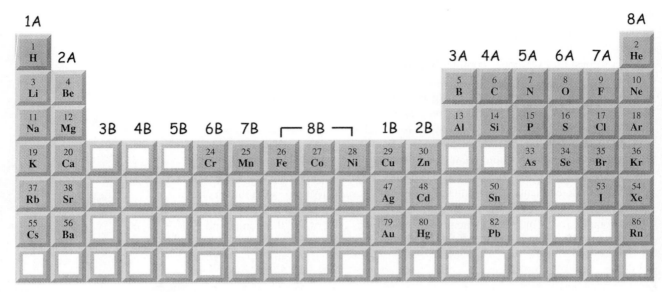

Figure 2.3   You should memorize the names and symbols of these elements.

hanging in your classroom), so there is no need to memorize the exact location of the elements on the chart. *(Check out the website if you're interested in the origin of the names of the elements. It may help you remember the symbols for some of the elements and learn some interesting history about their discoveries.)*

As mentioned previously, there are basically two types of chemical compounds—molecular and ionic. Because there are straightforward, but distinct, rules for naming each of the two types, you first have to be able to distinguish between them based solely on the chemical formula. As you saw earlier, molecules are only made of nonmetals, whereas ionic compounds have both metal atoms and nonmetal atoms (with the notable exception of ammonium compounds). How do you know which atoms are metals and which are nonmetals? The dark zigzag line in the periodic table separates the metals (on the left) from the nonmetals (on the right). If you ever happen to forget which side is which, you can rely on what you already know from practical experience. Elements like oxygen and nitrogen, which you know are not metals (metals are not something you'd want to breathe in), are on the right, and the elements on the left side include some obvious metals like sodium, aluminum, and copper.

Naming Molecular Compounds

Most **molecular compounds** are composed of nonmetallic elements. For now, let's focus on compounds containing only two elements. We place the name of the first element in the formula first, and the second element is named by adding "-ide" to the root of the element name. For example, HCl is called hydrogen chloride, and HBr is called hydrogen bromide. Often we find that one pair of elements can form several different compounds. In these cases, confusion is avoided by using Greek prefixes to denote the number of atoms of each element present:

CO (carbon monoxide)

CO_2 (carbon dioxide)

SO_2 (sulfur dioxide)

SO_3 (sulfur trioxide)

NO$_2$ (nitrogen dioxide)

N$_2$O$_4$ (dinitrogen tetraoxide)

Here is how Democritus would have counted from one to ten: 1 (*mono*), 2 (*di*), 3 (*tri*), 4 (*tetra*), 5 (*penta*), 6 (*hexa*), 7 (*hepta*), 8 (*octa*), 9 (*nona*), and 10 (*deca*). Have you noticed something odd about the first five compounds shown in the previous list? We did not use the prefix "mono-" to indicate the number of atoms present for the first element. This is done for convenience only because using a name like "monocarbon monoxide" to describe CO can be a mouthful. So the absence of the prefix for the first element means there is only one atom present.

2.4

(a) Name the compound SF$_6$, and (b) write the chemical formula of tetraphosphorus decaoxide.

Answer

(a) *There are six fluorine (F) atoms and one sulfur (S) atom present. Using the Greek prefixes, we call the compound sulfur hexafluoride.*

(b) *Tetra means 4 and deca means 10. There are four phosphorus (P) and ten oxygen (O) atoms in this compound. Its chemical formula is P$_4$O$_{10}$.*

For naming molecular compounds, that's all there is to it—just use Greek prefixes. *(Check out the problems on the website, and you'll be convinced.)*

EXAMPLE

Naming Ionic Compounds

You have already seen some of the rules for naming ionic compounds, which are made up of cations (positive ions) and anions (negative ions). As we said earlier, the key to naming ionic compounds is knowing the charge on the ions, and the periodic table can be used easily to determine the charge on monatomic ions of the elements in Groups 1A–7A (the so-called **representative elements**). Most of these elements will either lose or gain electrons (1, 2, or 3) in order to have the same number of electrons as a noble gas, which represents stability. (Remember the general rule: metals lose electrons, and nonmetals gain electrons.) Thus, all of the Group 1A elements lose one electron, the Group 2A elements lose two, the Group 3A metal (Al is the only important one) loses three, the Group 5A nonmetals gain three, the Group 6A elements gain two, and the Group 7A elements gain one. For example, Na will lose one electron to become Na$^+$, which has the same number of electrons as the noble gas Ne, while Cl will gain one electron to become Cl$^-$, which has the same number of electrons as the noble gas Ar.

To name the positive monatomic ions, simply add ion to the element's name. For example, Mg^{2+} is called magnesium ion. (By convention, we write the number before the $+$ or $-$ sign to indicate the charge on the ion.) To name negative monatomic ions, add '-ide' to the first part of the name. For example, N^{3-} is nitride, O^{2-} is oxide, and F$^-$ is fluoride. The "-ide" ending is also used for certain anion groups containing different elements, such as hydroxide (OH$^-$) and cyanide (CN$^-$). (Add CN$^-$ to you collection of flash cards.) The deadly poisonous compound KCN is called potassium cyanide.

In writing the chemical formula for an ionic compound, it is essential to include the appropriate numbers of cations and anions, indicated by a subscript, to ensure a neutral charge. For example, sodium carbonate contains sodium ions and carbonate ions, with charges of $+1$ and -2, respectively. In order to form a neutral species, there must be two sodium ions for every carbonate ion. This can be indicated by writing the formula as Na_2CO_3. Some other examples are $CaCl_2$, $Mg_3(PO_4)_2$, $(NH_4)_2SO_4$, BaO, and KNO_3. Naming ionic compounds follows from the formula in that the cation precedes the anions. Thus, the names of the examples just given are calcium chloride, magnesium phosphate, ammonium sulfate, barium oxide, and potassium nitrate. Other examples are lithium bromide ($LiBr$), calcium iodide (CaI_2), and aluminum oxide (Al_2O_3). We do not use Greek prefixes when naming ionic compounds because the subscripts can be inferred from the charges on the ions.

Many of the **transition metals** (Groups 1B, 3B–8B) can assume different charges (all positive), and their charge can be easily inferred from the number and charge of the anions present in the chemical formula. To indicate the charge in the name (for metals that can have different charges only), you have to include a Roman numeral in parenthesis after the cation's name. For example, Fe can become Fe^{2+} or Fe^{3+} to form compounds like $FeCl_2$ or $FeCl_3$; the names of these compounds are iron(II) chloride and iron(III) choride, respectively.

2.5 EXAMPLE

(a) Write the formula of the compound calcium phosphate, and (b) give the name of the compound $Ba(NO_3)_2$.

Answer

(a) *The calcium ion bears a $+2$ charge, and the phosphate ion bears a -3 charge. To balance the charges for electrical neutrality, we need to have three Ca^{2+} ions and two PO_4^{3-} ions ($2 \times 3 = 3 \times 2$). Therefore, the formula is $Ca_3(PO_4)_2$.*

(b) *The cation is barium (Ba^{2+}), and the anion is nitrate (NO_3^-), so the name is barium nitrate.*

For naming ionic compounds, that's all there is to it, and you can check out the problems on the website to convince yourself. Unfortunately, the names of the polyatomic ions cannot be inferred from the periodic table, and you will have to do a bit more memorization to get these down.

Naming Acids and Polyatomic Ions

For many students, the hardest part of learning much of the chemical nomenclature in general chemistry is keeping the polyatomic ions straight. You'll be expected to know at least 30 of them with different numbers and types of atoms as well as charges (although they are almost all anions). You could memorize them all, but it's easier to remember most of them if you know where they come from (that is, how they're produced). They come from dissolving acids in water.

You are no doubt familiar with the term **acid** (which is a substance that produces H^+ ions when dissolved in water), and you've had some practical experience with the properties of acids, like citric acid in lemon juice or the acetic acid in vinegar. Citric acid in water breaks down to form two ions, a hydrogen ion, H^+, and the polyatomic anion citrate; acetic acid breaks down in water to form a hydrogen ion, H^+, and the polyatomic anion acetate. Most acids dissolve in water to form hydrogen cations and polyatomic anions (although some, such as hydrochloric acid, HCl, form monatomic anions like Cl^-, which you already know how to name). If you've been reading carefully, you are probably thinking, "But you told me earlier that HCl is called hydrogen choride." As it turns out, the molecule HCl is a colorless gas, and it's called hydrogen chloride. When hydrogen chloride is bubbled into water, the molecule dissolves and ionizes as an acid into H^+ and Cl^- ions. For this reason, we call the dissolved HCl, represented as $HCl(aq)$, hydrochloric acid.

In order to use the acids to help remember the names of polyatomic anions, you will first have to memorize the chemical formulas of a handful of common acids. Let's start with four reference acids from which many others can be derived. These are nitric (HNO_3), phosphoric (H_3PO_4), sulfuric (H_2SO_4), and chloric ($HClO_3$). These acids are called **oxoacids** because they contain oxygen. They all share the common features in that they contain one or more H and O atoms and one central atom. If we remove an O atom from the acid, we change the name by converting "-ic" to "-ous." Thus, when HNO_3 becomes HNO_2, it is called nitrous acid; H_3PO_3 from H_3PO_4 is called phosphorous acid; H_2SO_3 from H_2SO_4 is called sulfurous acid; and $HClO_2$ from $HClO_3$ is called chlorous acid. Things are a bit more complicated for the oxoacids containing the halogens (Cl, Br, and I). It turns out that we can remove yet another O atom from chlorous acid to get HClO, which is called hypochlorous acid. Furthermore, we can also add an O atom to chloric acid to get $HClO_4$, which is called perchloric acid. We can now summarize all these names (which you should add to your collection of flash cards) in the following list:

$HClO_4$ (perchloric)

HNO_3 (nitric)	H_3PO_4 (phosphoric)	H_2SO_4 (sulfuric)	$HClO_3$ (chloric)
HNO_2 (nitrous)	H_3PO_3 (phosphorus)	H_2SO_3 (sulfurous)	$HClO_2$ (chlorous)

$HClO$ (hypochlorous)

You may also encounter the oxoacids of the halogens bromine and iodine, but these are easy to name if you know the four oxoacids of chlorine. For example, the fully oxygenated acids of bromine and iodine are perbromic acid and periodic acid. Two additional acids that are commonly seen in general chemistry are carbonic acid (H_2CO_3) and acetic acid ($HC_2H_3O_2$). Note that for acetic acid, only one of the four hydrogens is ionizable, so it is written at the beginning of the formula. The other common acids that do not contain O atoms are hydrofluoric acid (HF), hydrochloric acid (HCl), hydrobromic acid (HBr), hydroiodic acid (HI), and hydrocyanic acid (HCN). The last acid produces the cyanide ion (CN^-) when dissolved in water. Add these seven acids to your collection of flash cards.

If you count them, the total number of acids you should remember is about 20 (but learning the four reference acids first will make memorizing many of them pretty straightforward). The only acids you will ever come in contact with in a general chemistry laboratory are just five: nitric acid, phosphoric acid, sulfuric acid, hydrochloric acid, and acetic acid. Then why bother to learn the names of other acids? Because they are important in research and are often used as examples in problems. So knowing their names, structures, and properties will help you in problem solving.

Once we have tackled the names of acids, it's relatively easy to learn the names of anions derived from them. Anions derived from oxoacids are called **oxoanions.** The following rules are used to name oxoanions.

- When all the H ions are removed from the "-ic" acid, the anion's name ends with "-ate." For example, the anion SO_4^{2-} derived from H_2SO_4 is called sulfate.

- When all the H ions are removed from the "-ous" acid, the anion's name ends with "-ite." Thus, the anion NO_2^- derived from HNO_2, is called nitrite.

- The names of anions in which one or more but not all the H atoms have been removed must indicate the number of H atoms present. For example, if an H atom is removed from H_3PO_4 as H^+ ion, we get $H_2PO_4^-$, which is called dihydrogen phosphate.

It might help you to memorize these rules if you make a few more flash cards with some simple reminders such as 'ic' → 'ate' and 'ous' → 'ite'. Now we can draw a corresponding list of oxoanions like the oxoacids shown earlier (and you can add these to your collection of flash cards):

NO_3^- (nitrate)	$H_2PO_4^-$ (dihydrogen phosphate)	HSO_4^- (hydrogen sulfate)	ClO_4^- (perchlorate)
NO_2^- (nitrite)	HPO_4^{2-} (hydrogen phosphate)	SO_4^{2-} (sulfate)	ClO_3^- (chlorate)
	PO_4^{3-} (phosphate)	HSO_3^- (hydrogen sulfite)	ClO_2^- (chlorite)
			ClO^- (hypochlorite)
		SO_3^{2-} (sulfite)	

This list (and your flash card collection) can be expanded by adding the oxoanions of bromine and iodine (which are named like those of chlorine), carbonate (CO_3^{2-}), hydrogen carbonate (HCO_3^-), acetate ($C_2H_3O_2^-$).

We have seen that some acids, like HCl and HNO_3, can lose one H^+ ion per molecule, while others like H_2SO_4 can lose up to two H^+ ions. So HNO_3 is an example of **monoprotic acid** (one proton donor), and H_2SO_4 is a **diprotic acid** (two proton donor). Because H_3PO_4 can lose up to three protons, it is called a **triprotic acid.**

Now we come to bases. A **base** is a substance that produces one or more hydroxide ions (OH^-) when dissolved in water. Compared to acids, the naming of bases is very simple because there are just a few of them. In fact, in the general chemistry laboratory, the only bases you will ever work with are sodium hydroxide (NaOH) and ammonia (NH_3). Other bases that you will come across in problems and exams are the hydroxides of Group 1A elements (the alkali metals), such as LiOH and KOH, and the hydroxides of Group 2A elements (the alkaline earth metals), such as $Mg(OH)_2$, $Ca(OH)_2$, and $Ba(OH)_2$.

Once you've mastered the nomenclature rules described, you may want to be familiar with some of the old-fashioned, but still frequently used names:

Ferric ion	Fe^{3+}
Ferrous ion	Fe^{2+}
Cupric ion	Cu^{2+}
Cuprous ion	Cu^+
Bicarbonate	HCO_3^-
Bisulfate	HSO_4^-

Physical Properties of Compounds

You probably already know a lot about the physical properties of compounds based on your experience with the compounds you eat every day.

Now that you can distinguish molecules from ionic compounds (how?) and name them, let's start thinking a little about some of their basic physical properties using your own practical experience as a guide. We can tell if a compound is ionic or molecular by looking at its chemical formula; if it contains metals and nometals, it's likely to be ionic; if not, it's a molecule. This is really helpful because knowing whether a compound is ionic or molecular tells us a lot about its **physical properties.** (It also tells us a lot about its **chemical properties,** but we'll discuss those in later chapters.) Physical properties refer to those that can be studied without changing the identity of the substance. For example, if we say that gold has a bright yellow metallic luster, we are referring to the gold's physical property because the metal remains unchanged after we made the observation. Likewise, the melting point, boiling point, and densities of a substance all refer to its physical properties. If we boil water, do we change the identity of the molecule? No, it's still H_2O, but the molecules are no longer touching each other.

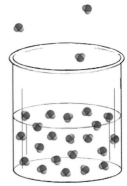

Boiling water does not change the identity of water molecules.

The melting point and boiling point of a compound can often be determined with ease, and they tell us quite a bit about the forces holding the units (molecules or ions) together. Ionic compounds generally have much higher melting points and boiling points than molecular compounds because the electrostatic forces between cations and anions are much stronger than those holding molecules together. **Note that we are comparing the forces that hold ions together in an ionic compound with the forces holding molecules together in the liquid or solid state, not the chemical bonds within a molecule that hold the atoms together.** For example, the melting point of sodium chloride (NaCl) is 801°C, whereas that of glucose ($C_6H_{12}O_6$), blood sugar, is only 146°C. That is, it takes more heat to separate Na^+ and Cl^- ions than to separate glucose molecules from one another. Actually, the melting points of molecular compounds vary all over the map—at room temperature, some are solids (sugar), some are liquids (water), and some are gases (oxygen). *All* ionic compounds are solids at room temperature. You should have a pretty good sense for the high melting and boiling points of ionic compounds if you imagine trying to melt or boil table salt (NaCl). Most ionic compounds are also called salts, although they come in different colors depending on the elements involved.

Another property that we can study is **solubility,** with which you have probably had lots of experience. If we add a compound to water, will it dissolve; that is, will it disappear? If so, then water acts as the **solvent,** the added compound is called the **solute,** and the resulting mixture is called the **solution.** The more a compound dissolves, the greater is the solubility of the compound. Unfortunately, there is no general rule that helps us predict the solubility of a substance, whether it is molecular or ionic (but it's very easy to find out either by doing the experiment or by looking it up in a handbook). However, we can investigate the resulting solution and learn something about the solute. When an ionic compound dissolves in water, the ions break loose from one another, and the resulting solution becomes electrically conducting. This is so because the movement of the cations and anions (in opposite directions) is similar to the movement of electrons along a piece of copper wire. For this reason, we call the compound an **electrolyte.** The vast majority of molecular compounds lack this property and are called **nonelectrolytes.** (Why?) This means that water solutions of these compounds cannot conduct electricity.

Molecular compound
in water

Ionic compound in
water

At this point, you may wonder that if ions are held so strongly in a solid, what causes them to break apart when we pour salt into our soup? The answer has to do with the nature of water as a solvent. Although water is not made up of ions, it does have a positive end (at the H atoms) and a negative end (at the O atom), so the overall charge of the molecule is zero. For this reason, water is called a **polar molecule** because it has the polarities of positive and negative charges. When sodium chloride is added to water, the ions break loose and become surrounded by the polar water molecules as shown in Figure 2.4. This interaction stabilizes the cations and anions individually and prevents their aggregation.

Why does the water molecule have positive and negative ends? Because the properties of elements gradually change from metallic (e.g., want to lose electrons) to nonmetallic (e.g., want to gain electrons) as we move across the periodic table, you might expect the 'desire' to gain electrons (or affinity for electrons) would increase as we move from left to right on the periodic table. Moreover, you might expect that because the zigzag line is slanted to the right, the nonmetallic properties would increase toward the top right corner of the table. In fact, this is true, and the atom with the strongest 'desire' or affinity for electrons is F, followed by O, N, and Cl.

Now think about what happens when O, with a high affinity for electrons, forms a chemical bond (i.e., shares electrons) with H, which has a lower affinity for electrons. The sharing is not likely to be equal (sort of like the sharing of toys between a 10-year-old boy and his 2-year-old brother). Oxygen (the bigger boy) will grab the lion's share of the negatively charged electrons in the chemical bond, imparting a

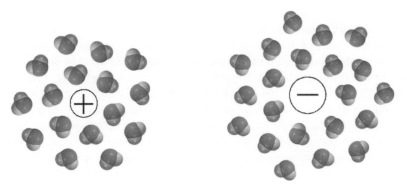

Figure 2.4 Water molecules surround ions and interact with them through ion-dipole intermolecular forces (to be discussed later in the book).

partial negative charge to the oxygen end of the molecule. Because water is neutral, the hydrogen end would have to be correspondingly positive. The unequal sharing of electrons is the basis for polarity in molecules.

Chemical compounds can be classified as either ionic or molecular (and you now know how to tell the difference). You should also know that molecules can be either polar or nonpolar. In fact, all compounds can be classified as either ionic, polar, or nonpolar. For now, it would be useful for you to remember these general rules. Ionic compounds and polar molecules tend to dissolve in polar solvents, while nonpolar (i.e., oily) molecules dissolve in nonpolar solvents.

You already have a lot of practical experience with this general rule of solubility. You know that oil and water don't mix, and you have certainly observed NaCl or the polar sugar molecule dissolving in water. What do you think would happen if you tried to dissolve NaCl in oil or gasoline?

2.6

EXAMPLE

Water is a good solvent for ionic compounds but is a nonelectrolyte, like most molecular compounds. Why, then, are we cautioned not to handle electrical appliances when our hands are wet with tap water?

Answer   *It is true that water is a nonelectrolyte and therefore cannot conduct electricity. But this description applies to pure water such as distilled water. Water from a tap comes originally from underground sources and usually contains enough dissolved ions to become electrically conducting.*

Test your understanding of the material in this chapter

IMPORTANT TERMS

Explain the following terms in your own words:

CONCEPTS

Explain in your own words:

- The structure of an atom.
- The difference between metals and nonmetals.
- The difference between molecules and ionic compounds and how to distinguish between the two based on their chemical formulas.
- The physical properties of boiling point, melting point, and solubility and how these properties differ among molecular and ionic compounds.

UNDERSTANDING CHEMISTRY

To test your overall understanding of the material in this chapter, use what you have learned to answer this question: What comes to mind when you look at a chemical formula? For example, consider the chemical formula Na_2CO_3.

You should immediately note that the compound is made of a metal and two nonmetals so it is probably ionic (and named accordingly). Each of the Na atoms would have lost an electron to become a sodium ion, a cation, and the rest of the compound must be the anion. You should recognize the anion as carbonate with a charge of -2. Thus, the name of this ionic compound is sodium carbonate. Because it's ionic, it must be a solid at room temperature and have a high melting point and boiling point. You might also expect it to dissolve in a polar solvent such as water. If all this comes to mind when you read Na_2CO_3, you are understanding chemistry. A useful mental exercise to aid in your understanding of chemistry would be to go through a similar analysis whenever you look at a chemical formula. This will help you think about chemicals in terms of their structure and properties and not just as an abstract string of letters.

Chemical Reactions
Changing partners

CHAPTER 3

The Chinese words for chemistry are "study of change," which quite accurately describes the subject. As we saw in Chapter 2, atoms undergo chemical changes when they react to form molecules and ions. Of course, molecules and ions also react with one another to form other molecules and/or ions. Although the total number of known elements is slightly over 100, the number of ways that atoms, molecules, and ions can react with one another is in the tens of millions (but don't panic—most of these reactions are just variations on a few types). Before we discuss types of chemical changes, you should learn how to describe these changes using **chemical equations.** Once you are comfortable with representing chemical reactions with chemical equations, you will learn how to recognize and think about a few important types of chemical changes.

Chemical Equations

In order to write about chemical reactions, chemists have devised a standard way to represent them. Consider what happens when hydrogen gas (H_2) burns in air (which contains oxygen, O_2) to form water (H_2O). This reaction can be represented by the chemical equation

$$H_2(g) + O_2(g) \longrightarrow H_2O(l)$$

where the + sign means "reacts with" and the $\longrightarrow$ sign means "to yield" (that is, to produce). Thus, this equation can be read: "Molecular hydrogen gas reacts with molecular oxygen gas to yield water." The reaction is assumed to proceed from left to right as the arrow indicates. But there's a problem with the equation as it is written (do you see what it is?). We have two oxygens on the left side of the arrow and only one oxygen on the right. Chemists would say that the equation is not balanced. Chemical equations must follow an extremely important rule (so important that it's given a "law" status in chemistry): you can't create matter (i.e., atoms), and you can't destroy it, which is one way of stating the **law of conservation of mass.** To abide by the law, we have to make sure that chemical equations have the same number of each type of atom on both sides of the arrow. That is, we must have as many atoms after the reaction ends as we did

The Chinese characters for chemistry mean "the study of change."

before it started. We can balance this expression by placing an appropriate coefficient (2 in this case) in front of H_2 and H_2O:

$$2H_2 + O_2 \longrightarrow 2H_2O$$

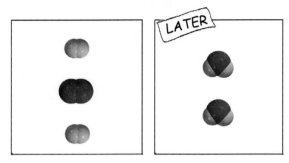

Two H_2 molecules collide with one O_2 molecule, and two H_2O molecules are made.

This balanced chemical equation shows that "two hydrogen molecules can combine or react with one oxygen molecule to form two water molecules." H_2 and O_2 in the previous equation are the **reactants,** the starting materials in the reaction, and water is the **product,** the substance formed as a result of the reaction. A chemical equation, then, can be thought of as the chemist's shorthand description of a reaction. (As we will see later, a chemical equation is quite similar to an algebraic equation like $a \times b = c$.) In a chemical equation, the reactants are conventionally written on the left and the products on the right of the arrow:

$$\text{reactants} \longrightarrow \text{products}$$

To accurately describe a reaction, it is important to use the abbreviations g, l, s, or aq in parentheses to describe the physical state of the chemical—gas, liquid, solid, or aqueous (which means dissolved in water, as shown in Figure 2.4). For example,

$$2CO(g) + O_2(g) \longrightarrow 2CO_2(g)$$

$$2HgO(s) \longrightarrow 2Hg(l) + O_2(g)$$

To describe the processes of sodium chloride dissolving in water, we write

$$NaCl(s) \xrightarrow{\text{H}_2\text{O}} NaCl(aq)$$

Writing H_2O above the arrow symbolizes the physical process of dissolving a substance in water (which is not a chemical change), although it is sometimes left out for simplicity.

These abbreviations not only remind us of the physical state of the reactants and products, but they also tell us how to carry out experiments involving the chemical reactions. For example, when aqueous solutions of potassium bromide (KBr) and silver nitrate ($AgNO_3$) are mixed together, a solid, silver bromide (AgBr), is formed and settles at the bottom of the container. This reaction can be represented by the equation

$$KBr(aq) + AgNO_3(aq) \longrightarrow KNO_3(aq) + AgBr(s)$$

The same equation omitting the physical states of reactants and products would be

$$KBr + AgNO_3 \longrightarrow KNO_3 + AgBr$$

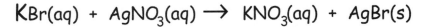

$$KBr(aq) + AgNO_3(aq) \rightarrow KNO_3(aq) + AgBr(s)$$

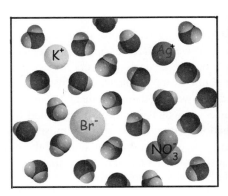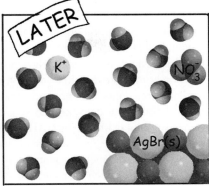

Dissolved silver ions and bromide ions
combine to form insoluble silver bromide.

Because no physical states are given, one might try to carry out this reaction by mixing solid potassium bromide with solid silver nitrate. But solid potassium bromide and silver nitrate, when mixed together, would react very slowly or not at all. If we could look at the process on the microscopic level, we'd see that when dissolved, Ag^+ and Br^- ions (from the ionic compounds $AgNO_3$ and KBr) bump into each other to form $AgBr$. In the solid state, these ions are locked in place and have little mobility.

3.1

EXAMPLE

When iron (Fe) is exposed to moist air (which contains oxygen), it slowly forms rust, which is Fe_2O_3. The same compound can be formed more rapidly by burning iron in an atmosphere of oxygen gas. Describe in words what can be deduced from the following chemical equation:

$$4Fe(s) + 3O_2(g) \longrightarrow 2Fe_2O_3(s)$$

Answer *1. Solid iron reacts with O_2 (oxygen gas) to form solid iron oxide, Fe_2O_3 (what is this called?). 2. Four iron atoms react with three oxygen molecules to produce two units of Fe_2O_3.*

A chemical equation is simple and informative in describing the overall change, but it's not a complete description of what actually happens during the reaction. It says nothing about how products are formed from reactants or how long it will take for the change to occur. The reaction discussed in Example 3.1 is a good illustration of the limitation of chemical equations. Normally, the rusting of an iron nail would take days or weeks. On the other hand, Fe_2O_3 can be formed within seconds when iron is burned in oxygen. Yet, both of these processes are represented by the same equation.

Balancing Chemical Equations

Suppose we want to write an equation to describe a chemical reaction that we have just carried out in the laboratory. How should we go about doing this? Because we know the identities of the reactants, we can write the chemical formulas for them. The identities of products are more difficult to establish. For simple reactions, it is often possible to guess the product(s). For more complicated reactions, where there may be three or more products, chemists may need to carry out further tests to figure out the identity of the products. As we will see later in this chapter, knowing the type of reaction will often help us guess intelligently what products may be formed. We can also learn a lot about a reaction by *observing* its progress. For example, we can conclude that a gaseous product is formed if we see bubbles appearing in an aqueous reaction. Color change is another indication that a chemical reaction has occurred and can sometimes provide clues about the nature of the products. Once we have identified all the reactants and products and have written their chemical formulas, we assemble them in the conventional sequence—reactants on the left separated by an arrow from products on the right. The equation written at this point is most likely to be *unbalanced,* that is, the number of each type of atom on both sides of the equation is not equal. Balancing equations is just a matter of making sure that we have the *same* number of each type of atoms on both sides of the arrow.

Let's consider an example. Methane (CH_4), the gas used in Bunsen burners and gas cooking-ranges, burns in air (which contains oxygen) to form carbon dioxide (CO_2) and water (for convenience, we omit the physical states of the chemicals here).

$$CH_4 + O_2 \longrightarrow CO_2 + H_2O$$

First we note that the C atoms are already balanced, but the H and O atoms are not. We can deal with either one first; it really doesn't matter. To balance the H atoms, we put a "2" in front of H_2O and get

$$CH_4 + O_2 \longrightarrow CO_2 + 2H_2O$$

All that remains is to put a "2" in front of O_2 (to balance the O atoms). The final balanced equation now looks like

$$CH_4 + 2O_2 \longrightarrow CO_2 + 2H_2O$$

A quick tally shows that there are one C atom, four H atoms, and four O atoms on both sides of the arrow.

3.2

EXAMPLE

Write a balanced equation to represent the reaction between sodium metal (Na) and water to form sodium hydroxide (NaOH) and hydrogen gas (H_2).

Answer *First we write*

$$Na + H_2O \longrightarrow NaOH + H_2$$

We see that both the Na and O atoms are balanced, but the H atoms are not. Because the H atoms will always be even on the left of the arrow, we multiply NaOH by 2 so that the H atoms are also even on the right of the arrow and obtain

$$Na + H_2O \longrightarrow 2NaOH + H_2$$

All that is left is to add a "2" to both Na and H_2O to arrive at the final equation:

$$2Na + 2H_2O \longrightarrow 2NaOH + H_2$$

The final tally shows that there are two Na atoms, four H atoms, and two O atoms on both sides of the arrow.

Visit the website for more problems on balancing chemical equations.

Major Types of Chemical Reactions | *There are only three!*

Nearly all of the chemical reactions you will see in general chemistry are one of three types: precipitation, acid-base, and oxidation-reduction. In fact, **most chemical reactions are either acid-base or oxidation-reduction,** which, as we will see later, involve the transfer of protons or electrons, respectively. Imagine that! Most of the millions of chemical reactions involve the transfer of only one or the other of these subatomic particles. If you keep this in mind, it should prove to be another bonus for understanding chemistry.

Precipitation Reactions

Precipitation reactions usually occur in aqueous solution and result in the formation of an insoluble product, or **precipitate.** (In chemistry, precipitation refers to the formation of a solid from an aqueous or liquid environment as opposed to the precipitation that weather forecasters talk about when water vapor condenses to form rain.) Most of the precipitation reactions you will see in general chemistry involve ionic compounds. For example, when an aqueous solution of lead nitrate [$Pb(NO_3)_2$] is mixed with an aqueous solution of potassium iodide (KI), a yellow precipitate of lead iodide (PbI_2) is formed:

$$Pb(NO_3)_2(aq) + 2KI(aq) \longrightarrow PbI_2(s) + 2KNO_3(aq)$$

This balanced equation is essential if we want to actually carry out the reaction because it tells us that we have to prepare solutions of lead nitrate and potassium iodide and then mix them together. But it doesn't effectively depict the reaction because the formation of lead iodide requires only the interaction between lead and iodide ions. There are three steps to writing an equation representing the precipitation of lead iodide (or any other similar reaction) in a way that shows only the participants in the reaction. The first step is to write the balanced chemical equation including all the reactants and products as shown previously. Because the reactants are aqueous solutions, we know that both $Pb(NO_3)_2$ and KI are dissociated into ions in water:

$$Pb(NO_3)_2(s) \xrightarrow{\text{H}_2\text{O}} Pb^{2+}(aq) + 2NO_3^-(aq)$$

and
$$KI(s) \xrightarrow{\text{H}_2\text{O}} K^+(aq) + I^-(aq)$$

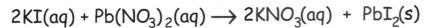

$$2KI(aq) + Pb(NO_3)_2(aq) \longrightarrow 2KNO_3(aq) + PbI_2(s)$$

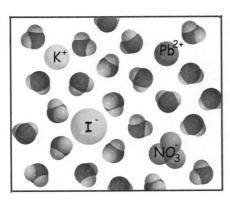 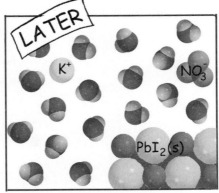

Dissolved lead ions and iodide ions combine to form insoluble lead iodide.

You can imagine these ions moving randomly in the mixed solution, and the encounter between Pb^{2+} and I^- ions leads to the build up of PbI_2 precipitate. The second step is to describe the reaction in terms of the dissociated ions and the solid product:

$$Pb^{2+}(aq) + 2NO_3^-(aq) + 2K^+(aq) + 2I^-(aq) \longrightarrow$$
$$PbI_2(s) + 2K^+(aq) + 2NO_3^-(aq)$$

This equation is called the **ionic equation** because the reactants and products are shown as ions. Note that K^+ and NO_3^- are present on both sides of the equation. This implies that nothing happens when K^+ ions bump into NO_3^- ions; that is, these ions don't participate in the reaction. For this reason, they are called *spectator ions* (very much like what you do when you go to see a basketball game). Note that PbI_2 is not written as ions because it is a solid.

Because a chemical equation is like an algebraic equation, we can cancel the K^+ and NO_3^- ions on both sides of the arrow. Therefore, the third and last step of representing this reaction is to show what is actually taking place in solution:

$$Pb^{2+}(aq) + 2I^-(aq) \longrightarrow PbI_2(s)$$

This is called the **net ionic equation** because it describes only the net change of the reaction.

In studying precipitation reactions, you are expected to be familiar with all three steps in writing equations. The first step is useful because it tells you which reagents to use. The second step, the total ionic equation, gives a more accurate picture of all the species present in the reaction mixture. Finally, the net ionic equation focuses only on the change that occurs in solution. As we will see shortly, net ionic equations are also helpful in following acid-base and oxidation-reduction reactions.

In the previous example, we were given the reactants and products and their physical states. But what if we were given only the starting materials? Can we tell beforehand if a precipitate will form when we mix two solutions together? Yes, because it is simply based on whether or not a compound is soluble, and the solubilities of thousands of compounds have been determined experimentally and are compiled in convenient tables. Of course, you don't have to memorize the solubilities of all these compounds if you can look them up in a table. But it's helpful to remember this: The presence of certain ions in a compound will make that compound soluble in water regardless of the nature of the counter ion in the compound. For example, the following generalizations are useful to keep in mind:

1. **Compounds containing the alkali metal ions (for example, Li^+, Na^+, K^+) are soluble regardless of what the anions are.**
2. **Compounds containing the ammonium ion (NH_4^+) are soluble regardless of what the anions are.**

3. **Compounds containing the nitrate (NO_3^-), bicarbonate (HCO_3^-), and chlorate (ClO_3^-) ions are soluble regardless what the cations are.**

As you might expect, there are also ions that usually form insoluble compounds, and you are probably already familiar with some common examples. Have you ever tried to dissolve a piece of chalk? **Compounds containing carbonate or phosphate (except those of alkali metal and ammonium ions) are insoluble.** For example, the ionic compound calcium carbonate ($CaCO_3$), which is the major ingredient in blackboard chalk, is insoluble in water. The magnificent stalactites and stalagmites you may have seen in caves are composed of calcium carbonate as well, and their formation is simply a precipitation reaction. Our bones are made up mostly of another insoluble compound, calcium phosphate [$Ca_3(PO_4)_2$]. What about compounds containing all the other types of ions? You can always look up the solubility of a compound in a table. There are handy chemical reference books (called handbooks) and websites that list the solubilities for countless compounds.

Acid-Base Reactions

Acids and bases are as familiar to us as aspirin and milk of magnesia. In addition to being the basis of many medicinal and household products, acid-base chemistry is important in industrial processes and essential in sustaining biological systems.

Acids are generally defined as substances that release H^+ ions in solution. (Note that the H^+ ion is commonly referred to as a proton. When an H atom loses its electron, all that's left is a proton in the nucleus.) They have a sour taste (examples are vinegar, which contains acetic acid, and lemon, which contains citric acid), and they turn blue litmus paper red. Bases are substances that produce OH^- ions in solution. Some of them have a bitter taste, and they turn red litmus paper blue.

The most practical definition of acids and bases was provided by the Danish chemist *Johannes Brønsted.* According to the definition, a **Brønsted acid** is a proton donor, and a **Brønsted base** is a proton acceptor. Within Brønsted's definition, therefore, an acid-base reaction can be viewed as a proton-transfer reaction, that is, from a Brønsted acid to a Brønsted base.

Let's apply this definition to hydrochloric acid—HCl(*aq*). In Chapter 2 we saw that in the gas phase, HCl is called hydrogen chloride and is a molecular compound. First, we bubble HCl gas into water so it is transferred to the aqueous environment:

AN ACID IS A PROTON DONOR AND A BASE IS A PROTON ACCEPTOR. ACCORDING TO MY DEFINITION, AN ACID-BASE REACTION INVOLVES THE TRANSFER OF A PROTON.

Brønsted defines acids and bases.

$$HCl(g) \xrightarrow{\text{H}_2\text{O}} HCl(aq)$$

Once in water, HCl ionizes into H^+ and Cl^- ions

$$HCl(aq) \longrightarrow H^+(aq) + Cl^-(aq)$$

and this ability to release H^+ in water qualifies HCl as a Brønsted acid. However, the Brønsted definition of an acid is a proton donor. To whom does the Brønsted acid HCl donate a proton? The only other species around is water; the proton in this case is transferred to H_2O. According to the Brønsted definition, water is acting as a base because it is accepting a proton.

A proton is extremely small with a diameter of about 10^{-15} m, compared to a diameter of 10^{-10} m for an average atom. To really understand acid-base chemistry, you need to think about what happens to a proton in water. This tiny charged particle is unlike any other positively charged species you might run into in an aqueous solution. As it turns out, the proton has such a strong attraction for the negative pole (the O atom) of the polar water molecule that it readily sticks to any water molecule in sight. In fact, it is known that the proton exists in an associated form with one or more H_2O molecules. For simplicity, we assume that it is only one water per proton and write the formula as $H^+(H_2O)$ or simply H_3O^+, which is called the **hydronium ion.** The ionization of hydrochloric acid can now be expressed more correctly as

$$HCl(aq) + H_2O(l) \longrightarrow H_3O^+(aq) + Cl^-(aq)$$

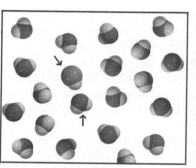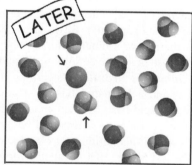

Hydrochloric acid donates a proton to water
to produce hydronium ion and chloride ion.

This is an example of an acid-base reaction in which HCl acts as a Brønsted acid to donate a proton (H^+) to water, which acts as a Brønsted base.

Acids that you will frequently encounter (carefully) in the laboratory are hydrochloric acid (HCl), nitric acid (HNO_3), acetic acid ($HC_2H_3O_2$), sulfuric acid (H_2SO_4), and phosphoric acid (H_3PO_4). The first three are examples of monoprotic acids; that is, each unit of the acid yields one hydrogen ion upon ionization:

$$HCl(aq) + H_2O(l) \longrightarrow H_3O^+(aq) + Cl^-(aq)$$

$$HNO_3(aq) + H_2O(l) \longrightarrow H_3O^+(aq) + NO_3^-(aq)$$

$$HC_2H_3O_2(aq) + H_2O(l) \rightleftharpoons H_3O^+(aq) + C_2H_3O_2^-(aq)$$

Notice that we used a single arrow for the first two reactions and a double arrow for the last one. The reason is that both HCl and HNO_3 are strong electrolytes, which means they have a tendency to completely split (or ionize) into cations and anions when dissolved in water. When the reaction is complete, essentially all of the acids have ionized. Because this is a one-direction process from left to right, we used a single arrow. The double arrows used for $HC_2H_3O_2$ mean this is a *reversible* reaction. Initially, a number of $HC_2H_3O_2$ molecules react with H_2O to form $C_2H_3O_2^-$ and H_3O^+ ions. As time goes on, some of the $C_2H_3O_2^-$ and H_3O^+ ions run into each other in solution and recombine to form $HC_2H_3O_2$ and H_2O molecules. At the same time, some other $HC_2H_3O_2$ molecules react with H_2O. Eventually, a state is reached in which the acid molecules ionize as fast as the ions recombine. Such a chemical state, in which no net change can be observed (although there is busy two-way traffic going on at the molecular level), is called **chemical equilibrium.** Chemical equilibrium is an extremely important concept we will discuss further in Chapter 8.

The main difference between HCl and HNO_3 on one hand and $HC_2H_3O_2$ on the other is that HCl and HNO_3 ionize completely and $HC_2H_3O_2$ does not. HCl ionizes completely because the H_3O^+ and Cl^- ions have no tendency to recombine to form HCl and H_2O. Likewise, HNO_3 ionizes completely because H_3O^+ and NO_3^- ions have no tendency to recombine. In contrast, H_3O^+ and $C_2H_3O_2^-$ ions do have a tendency to recombine to form acetic acid and water. So what really distinguishes the acids that ionize completely (the **strong acids**) from those that do not (the **weak acids**) is the anion. HCl is a strong acid because Cl^- simply does not want to hang on to a proton if it's in water; in other words, Cl^- is a very weak base because it's a poor proton acceptor. Similarly NO_3^- will not take a proton from H_3O^+ because it is also a very weak base. If you think about it, water must therefore be a stronger base (that is, a better proton acceptor) than either Cl^- or NO_3^- because all of the protons from the corresponding acids are associated with water. The proton will stick to whichever is the stronger base. On the other hand, only a small fraction of $HC_2H_3O_2$ molecules react to form $C_2H_3O_2^-$ and H_3O^+ ions and the rest of them remain intact. For example, in vinegar (that is, aqueous $HC_2H_3O_2$), only a few percent of the acetic acid molecules are ionized. Using the same reasoning as before, it follows that $C_2H_3O_2^-$ is a stronger base than either Cl^- or NO_3^-. Is $C_2H_3O_2^-$ a stronger base than water? Yes, because most of the protons end up associated with $C_2H_3O_2^-$ (as $HC_2H_3O_2$) instead of water.

Because HCl and HNO_3 ionize completely and are strong electrolytes, they are called strong acids, whereas $HC_2H_3O_2$ is called a weak acid because it ionizes partially and is a weak electrolyte. We will use a single arrow for strong acids and double arrows for weak acids. How do we know which acids are strong and which are weak? It's actually pretty easy because there are only six strong acids you will encounter in general chemistry. They are $H_2SO_4(aq)$, $HNO_3(aq)$, $HClO_4(aq)$, $HCl(aq)$, $HBr(aq)$, and $HI(aq)$. All the rest of the acids are weak. A general rule in writing chemical formulas of acids is that acids have hydrogen at the beginning of the

Weak acid   Strong Acid

chemical formula. At this point, you might wonder if water is an acid because it has hydrogen at the beginning of the formula. The answer is yes, but it's a very weak one. As we have seen, water can also act as a base. Water has the unusual property of being able to act as either an acid or a base.

Sulfuric acid (H_2SO_4) is a *diprotic acid* because each unit of the acid gives up two H^+ ions, which we can consider as two separate steps:

$$H_2SO_4(aq) + H_2O(l) \longrightarrow H_3O^+(aq) + HSO_4^-(aq)$$

$$HSO_4^-(aq) + H_2O(l) \rightleftharpoons H_3O^+(aq) + SO_4^{2-}(aq)$$

We use a single arrow for the first step because H_2SO_4 is a strong acid and double arrows for the second step because HSO_4^- (hydrogen sulfate) is a weak acid. Both H_2SO_4 and HSO_4^- behave as Brønsted acids because they are proton donors.

There are very few triprotic acids, acids that yield three H^+ ions when they ionize. The only one of importance in general chemistry is phosphoric acid (H_3PO_4), whose stepwise ionizations are

$$H_3PO_4(aq) + H_2O(l) \rightleftharpoons H_3O^+(aq) + H_2PO_4^-(aq)$$

$$H_2PO_4^-(aq) + H_2O(l) \rightleftharpoons H_3O^+(aq) + HPO_4^{2-}(aq)$$

$$HPO_4^{2-}(aq) + H_2O(l) \rightleftharpoons H_3O^+(aq) + PO_4^{3-}(aq)$$

Again, the double arrows tell us that all three species, H_3PO_4, $H_2PO_4^-$ (called dihydrogen phosphate), and HPO_4^{2-} (called hydrogen phosphate), are weak acids.

The Brønsted definition of a base is a substance that can accept a proton (H^+). We have already seen two examples of substances acting as bases—water and acetate ion. The most common base is the hydroxide ion (OH^-). Because the charged hydroxide ion cannot exist alone outside of water (that is, you won't ever see a bottle of pure hydroxide ions), it is usually found in ionic compounds like sodium hydroxide (NaOH), which is referred to as a **strong base.** Strictly speaking, NaOH itself is *not* a Brønsted base because it can't accept a proton. But NaOH dissociates completely in water to produces OH^- ions

$$NaOH(s) \xrightarrow{H_2O} Na^+(aq) + OH^-(aq)$$

and the OH^- ion can accept a proton. So it's the OH^- that is the base. Thus, if you add hydroxide ions to a solution containing an acid, they will accept protons from the acid to form water. Because an acid in water is really a solution containing $H_3O^+(aq)$ ions, it is more accurate to say that a hydroxide ion can react with a hydronium ion to produce two water molecules:

$$OH^-(aq) + H_3O^+(aq) \longrightarrow 2H_2O(l)$$

This equation is the net ionic equation of *all* acid-base reactions in the Brønsted scheme where OH^- is the Brønsted base and H_3O^+ is the Brønsted acid. At this point you may wonder if the OH^- ion would react with water itself. The answer is yes, and the reaction is

$$OH^-(aq) + H_2O(l) \rightleftharpoons H_2O(l) + OH^-(aq)$$

This is also a Brønsted acid-base reaction (with OH^- as the base and H_2O as the acid), but the products are the same as the reactants so we won't bother with it. Like diprotic acids, some bases contain two OH^- ions per unit, such as barium hydroxide:

$$Ba(OH)_2(s) \xrightarrow{H_2O} Ba^{2+}(aq) + 2OH^-(aq)$$

Note that like NaOH, $Ba(OH)_2$ is called a base even though it is the OH^- ions that act as the Brønsted base.

Another important base is ammonia (NH_3). How can it be a base if it doesn't contain an OH group like NaOH? Because in solution, ammonia will take a proton from a water molecule (which acts as an acid), according to the equation

$$NH_3(aq) + H_2O(l) \rightleftharpoons NH_4^+(aq) + OH^-(aq)$$

so it is a proton acceptor and therefore a Brønsted base. Note that we used the double arrows here because NH_3 is a **weak base** (and a weak electrolyte). In solution, very few NH_3 molecules undergo this reaction with water molecules. Most of them remain intact. (Although ammonia is a weak base, it's still a much stronger base than water.)

At this point we have discussed many different acids and only a few bases. Are there other bases? As it turns out, within every weak acid lies a weak base. Consider the example of acetic acid on p. 36. The double arrows in the equation indicate that the reverse reaction occurs when acetate ion takes a proton from hydronium ion; therefore, acetate ion acts as a base.

$$C_2H_3O_2^-(aq) + H_3O^+(aq) \longrightarrow HC_2H_3O_2(aq) + H_2O(l)$$

Similarly, all anions produced by the ionization of weak acids will accept a proton from hydronium ion. For example, SO_4^{2-}, $H_2PO_4^-$, HPO_4^-, and PO_4^{3-}, are all bases because they will accept protons from hydronium ions to form the parent acid as shown in the reverse reactions on p. 38. There is no need to memorize these or any other anions as long as you can recognize weak acids and you understand the nature of acid-base reactions.

Mixing Acids and Bases Can Produce Salts

When you mix an acid with a hydroxide compound in water, the protons from the acid react with the hydroxide ion to form water, and you're left with the cation from the hydroxide compound and the anion from the acid, or an aqueous solution of salt. Remember from Chapter 2 we said that most ionic compounds are **salts.** For example, if hydrochloric acid is added to a sodium hydroxide solution, the reaction that takes place is

$$HCl(aq) + NaOH(aq) \longrightarrow NaCl(aq) + H_2O(l)$$

Sodium chloride, then, is a salt that contains the cation from the base and the anion from the acid. Because both HCl and NaOH ionize completely in water, we can rewrite the previous equation as

$$H^+(aq) + Cl^-(aq) + Na^+(aq) + OH^-(aq) \longrightarrow Na^+(aq) + Cl^-(aq) + H_2O(l)$$

According to the equation, the only reaction that occurs is between H^+ and OH^- ions:

$$H^+(aq) + OH^-(aq) \longrightarrow H_2O(l)$$

Note that the H^+ ion is actually H_3O^+ in solution (see equation on p. 38).

Therefore, whenever H^+ and OH^- ions are produced by acids and bases in solution, they rapidly and completely combine to form H_2O molecules. This is the essence of an **acid-base neutralization reaction,** that is, the reaction between an acid and a base to produce a salt and water. The characteristic feature of such acid-base neutralization reactions is that the base is always OH^-.

The following are also examples of acid-base neutralization reactions:

$$HF(aq) + KOH(aq) \longrightarrow KF(aq) + H_2O(l)$$

$$H_2SO_4(aq) + 2NaOH(aq) \longrightarrow Na_2SO_4(aq) + 2H_2O(l)$$

At this point you may be feeling a bit overwhelmed by all these reactions, so it's important that you understand the chemistry and not get bogged down by the different chemical equations. There are a few important points to keep in mind that will help you understand all acid-base reactions.

1. **When you place an acid in water, the acid molecules will ionize to give protons to water to form hydronium ions, although the extent to which they do this varies. With strong acids, all of the molecules give up their protons, and with weak acids, only a small number of the molecules give up their protons to water.**

2. **The most common base, the hydroxide ion, can be produced in water by dissolving an ionic hydroxide compound such as NaOH. Weak bases, like ammonia or the anion from any weak acid (e.g., the acetate ion), will also produce hydroxide ions in water by taking a proton from water and producing a OH^- ion, but only to a small extent.**

3. **If you mix an acid solution with a hydroxide solution (e.g., NaOH), the protons from the acid will combine with the hydroxide ions to form water molecules. This is called an acid-base neutralization reaction, which is also characterized by the formation of a salt.**

For now, it is important that you understand these three points about acid-base chemistry. You should also be able to recognize an acid-base reaction from its chemical equation—it simply requires recognizing acids and bases themselves. The examples we've looked at in this chapter, as well as the acids described in Chapter 2 (with which you should be familiar by now), represent most of the ones you are likely to encounter in general chemistry.

Oxidation-Reduction Reactions

Whereas acid-base reactions can be characterized as proton-transfer processes, the last of the three classes of reactions, called **oxidation-reduction,** or **redox, reactions,** involves the transfer of electrons between two substances. One substance loses electrons, called **oxidation,** while the other substance gains electrons, called **reduction.** Which substances tend to lose electrons, and which tend to gain them? As we said in Chapter 2, nonmetals will grab electrons if given a chance, and metals like to lose electrons.

Consider the reaction between sodium metal and chlorine gas, both highly reactive elements. If we heat sodium in an atmosphere of chlorine, a violent reaction occurs. After the smoke has subsided, the only substance left is sodium chloride in the form of a white powder. Sodium chloride, as we know, is an ionic compound containing Na^+ and Cl^- ions. So in this reaction, the Na metal must have transferred electrons to the Cl_2 molecule. Although the electrons are transferred directly, it is often useful to think about the transfer as the sum of two separate processes, which can be written as **half-reactions:**

$$Na \longrightarrow Na^+ + e^-$$

$$Cl_2 + 2e^- \longrightarrow 2Cl^-$$

Note that because chlorine is a diatomic molecule, we need two electrons per molecule. In order for the electrons to add up, we have to multiply the sodium half-reaction by 2 to give the overall equation:

$$2Na(s) + Cl_2(g) \longrightarrow 2NaCl(s)$$

This reaction is an example of a redox reaction. The sodium metal, which has lost an electron, is said to be *oxidized,* and the chlorine molecules, which accept electrons, are said to be *reduced.* For this reason, Na is also called a **reducing agent** (because it reduces Cl_2 molecules), and Cl_2 is called an **oxidizing agent** (because it oxidizes the Na metal). This may seem confusing, but it's simply the result of the definitions of oxidation and reduction previously given. A useful mnemonic for redox reactions is OILRIG: **O**xidation **I**s **L**oss (of electrons) and **R**eduction **I**s **G**ain (of electrons).

There are many examples of redox reactions. Calcium metal reacts with oxygen to form calcium oxide (CaO):

$$2Ca(s) + O_2(g) \longrightarrow 2CaO(s)$$

In terms of half-reactions, we write

$$2Ca \longrightarrow 2Ca^{2+} + 4e^- \quad \text{oxidation}$$

$$O_2 + 4e^- \longrightarrow 2O^{2-} \quad \quad \text{reduction}$$

(Recall that Ca is a Group 2A element, so we expect each Ca atom to lose two electrons to form the Ca^{2+} ion.) Here, Ca is the reducing agent because it loses electrons, and O_2 is the oxidizing agent because it accepts electrons.

When we add zinc metal to a copper sulfate ($CuSO_4$) solution, we observe the disappearance of the blue color (due to Cu^{2+} ions in water) as a result of the reaction

$$Zn(s) + CuSO_4(aq) \longrightarrow ZnSO_4(aq) + Cu(s)$$

or, in terms of net ionic equation

$$Zn(s) + Cu^{2+}(aq) \longrightarrow Zn^{2+}(aq) + Cu(s)$$

Note that SO_4^{2-} is a spectator in this reaction. The electron-transfer process can be represented by the half-reactions

$$Zn \longrightarrow Zn^{2+} + 2e^-$$

$$Cu^{2+} + 2e^- \longrightarrow Cu$$

So far so good. However, it turns out that the vast majority of redox reactions do not involve the *complete* transfer of electrons. Consider, for example, the reaction between hydrogen and chlorine gases:

$$H_2(g) + Cl_2(g) \longrightarrow 2HCl(g)$$

The product, hydrogen chloride (HCl), is a molecular compound and does not contain any ions. Yet, this reaction is also classified as a redox reaction. Why? It's because the electrons in the chemical bond joining the two atoms, H—Cl, are shifted toward the Cl atom, which has a greater attraction for them. In this sense we can say that hydrogen is oxidized and chlorine is reduced even though there is only a partial transfer of electrons from hydrogen to chlorine. Essentially anytime you produce a polar molecule, where electrons are not shared equally, you are transferring electrons (if only a little).

To keep track of electrons in redox reactions in general, we use the term **oxidation number,** which is the number of charges the atom would have in a molecule (or ionic compound) if electrons were transferred completely. In an ionic compound, the electrons are transferred completely, so the oxidation number is the same as the charge on the ion. It is important to realize that oxidation number is simply a useful bookkeeping tool even if the atoms don't transfer their electrons completely. The element that shows an increase in oxidation number is oxidized, and the one that shows a decrease in oxidation number is reduced. Referring to the formation of sodium chloride, we can now write

$$\overset{0}{2\text{Na}} + \overset{0}{\text{Cl}_2} \longrightarrow \overset{+1\ -1}{2\text{NaCl}}$$

The numbers above the element symbols are the oxidation numbers. Because there is no charge on the Na and Cl atoms in their elemental state, their oxidation number is zero. This is a rule for all atoms in the elements. In NaCl, sodium has a charge of $+1$ and therefore also an oxidation number $+1$. By the same token, chlorine has an oxidation number of -1 in NaCl because it bears a negative charge. Similarly, for the formation of HCl, we have

$$\overset{0}{\text{H}_2} + \overset{0}{\text{Cl}_2} \longrightarrow \overset{+1\ -1}{2\text{HCl}}$$

Thus, hydrogen has an oxidation number of $+1$ and chlorine an oxidation number of -1. When assigning oxidation numbers for molecular compounds, how do we know which atom gets the electrons? (Remember, we're just pretending a complete transfer occurs for bookkeeping purposes.) In this case, we know that Cl gets most of the electrons because it's farther to the right on the periodic table (remember as we move across horizontal periods the atoms want electrons more).

In studying redox reactions, our first step is often assigning oxidation numbers to elements because it helps us identify the oxidizing and reducing agents. It is useful to remember the following rules:

1. **All elements have an oxidation number of zero.**
2. **The oxidation number of metal ions is just their charge.**
3. **Hydrogen tends to lose its electron to nonmetals and take electrons from metals. It usually has an oxidation number of $+1$ except when it is combined with a metal, for example, sodium hydride (NaH). Then its oxidation number becomes -1.**
4. **For the remaining elements (mostly the nonmetals), the oxidation number depends on with whom they bond. The electrons will be assigned to the atom that has the greater attraction for electrons, but let's just consider C, N, O, F, and Cl. The tendency for the element to accept electron(s) in a bond increases as follows: $C < N \approx Cl < O < F$. Fluorine *always* has an oxidation number of -1, while others may have either positive or negative oxidation numbers, depending on the particular compound (that is, with**

what they bond). For example, O will have only a positive oxidation number when it bonds with F. It has a negative oxidation number when it bonds with any other atom in a compound.

Here's a useful way to determine the oxidation number of an element. First, we need to know how atoms are connected to one another by chemical bonds. (You will gradually become familiar with chemical formulas showing chemical bonds as you learn the subject.) In ammonia the bonding scheme is

$$H \underset{|}{\overset{H}{-}} N - H$$

Each line represents a chemical bond made up of two electrons, one from each bonding atom (N and H). Because nitrogen has a greater tendency to accept electrons than hydrogen, we can illustrate this process as follows:

$$H - N - H$$

Each arrow represents the transfer of both electrons in the bond to the N atom. Remember, we're just bookkeeping here; the electrons aren't really transferred completely. Consequently, the N atom would have a net charge of -3 due to the three electrons from the H atoms and hence an oxidation number of -3, while each H atom will have a charge of $+1$ and an oxidation number of $+1$. Of course, we know that N doesn't really have a charge of -3 and hydrogen doesn't really have a charge of $+1$ because ammonia is a molecule, but the oxidation number tells us who gained the lion's share of the bonding electrons and who didn't.

In some molecules, atoms are joined by double bonds (two lines). An example is carbon dioxide (CO_2):

$$O = C = O$$

Because oxygen has a greater tendency to accept electrons than carbon, we have

$$O = C = O$$

In the end, each O atom will gain a net charge of -2 (oxidation number of -2), while the C atom will have a net charge of $+4$ (oxidation number $+4$).

Understanding oxidation numbers reinforces your understanding of chemical reactions and the fact that electrons are transferred completely to form ionic compounds or just partially to produce polar molecules.

Check out the website for exercises in classifying chemical reactions.

Test your understanding of the material in this chapter

IMPORTANT TERMS

Explain the following terms in your own words:

acid-base neutralization reaction, p. 39
acid-base reaction, p. 33
Brønsted acid, p. 35
Brønsted base, p. 35
chemical equation, p. 29
chemical equilibrium, p. 37
half-reactions, p. 40
hydronium ion, p. 36
ionic equation, p. 34
law of conservation of mass, p. 29
net ionic equation, p. 34
oxidation, p. 40
oxidation number, p. 42
oxidation-reduction reaction, p. 40
oxidizing agent, p. 41
precipitate, p. 33

CONCEPTS

Explain in your own words:

- The meaning of a chemical equation and how to balance it.
- The characteristics of a precipitation reaction and the usefulness of ionic and net ionic equations.
- Brønsted acids and Brønsted bases, weak and strong acids and bases, the hydronium ion, acid-base reactions, and salt.
- The nature of oxidation-reduction reactions, oxidizing and reducing agents, and oxidation number.

UNDERSTANDING CHEMISTRY

To test your understanding of the material in this chapter, particularly with respect to identifying the type of chemical reaction, you should become adept at recognizing chemical compounds. At this stage, you should know which are acids, which are bases, and which are salts.

Let us first focus on precipitation reactions. From the list of cations and anions that form soluble compounds, it is relatively easy to predict the outcome of a precipitation reaction. When a solution of barium nitrate [$Ba(NO_3)_2$] is mixed with a solution of sodium sulfate (Na_2SO_4), a precipitate is formed.

$$Ba(NO_3)_2(aq) + Na_2SO_4(aq) \longrightarrow \; ?$$

In solution, the starting substances are dissociated into Ba^{2+}, NO_3^-, Na^+, and SO_4^{2-} ions. We know that Na^+ and NO_3^- will not form a precipitate because both ions form soluble compounds. So the likely candidate for precipitation is $BaSO_4$. This is indeed the case. You may be interested to know that because $BaSO_4$ is insoluble and Ba is opaque to X ray, the compound is used medically to diagnose digestive tract ailments in a process called barium enema. As mentioned earlier, the most reliable source of the solubility of compounds is a chemistry handbook.

$BaSO_4$ is a salt in which the cation (Ba^{2+}) is derived from barium hydroxide [$Ba(OH)_2$] and the anion is derived from sulfuric acid (H_2SO_4):

$$\text{from } Ba(OH)_2 \; \nwarrow \; \overset{\displaystyle BaSO_4}{} \; \nearrow \; \text{from } H_2SO_4$$

So how would you prepare $BaSO_4$ by an acid-base reaction? Simply by reacting $Ba(OH)_2$ with the acid H_2SO_4:

$$Ba(OH)_2(aq) + H_2SO_4(aq) \longrightarrow BaSO_4(s) + 2H_2O(l)$$

Finally, let's see how the acid and the base used in these reactions are prepared. First consider $Ba(OH)_2$. Experiments show that when barium metal is placed in water, hydrogen gas is produced. The reaction can be represented by

$$Ba(s) + 2H_2O(l) \longrightarrow Ba(OH)_2(aq) + H_2(g)$$

Try to determine the oxidation numbers of the elements in both the reactants and products. Is it a redox reaction? Is it a precipitation reaction? If you answered yes to the first question and no to the second question, you are on the right track.

How about H_2SO_4? This is a more complicated case as it involves several steps. First, we burn sulfur in air to form sulfur dioxide:

$$S(s) + O_2(g) \longrightarrow SO_2(g)$$

Next, we react sulfur dioxide with oxygen to form sulfur trioxide:

$$2SO_2(g) + O_2(g) \longrightarrow 2SO_3(g)$$

Finally, we bubble the SO_3 gas into water to form sulfuric acid:

$$SO_3(g) + H_2O(l) \longrightarrow H_2SO_4(aq)$$

How would you characterize the first and second reaction?

4 Reactants to Products

Measure for measure

ow that you've learned how to interpret and balance chemical equations, you are ready to put this knowledge to practical use. Because chemical equations represent chemical reactions, they are used in practice to figure out how to prepare useful compounds. It's actually a lot like cooking. In order to prepare a certain amount of food, you have to use ingredients in the right proportions. In chemistry you have to use the right amount of reactants to yield a certain amount of product. In fact, a lot of cooking involves chemical reactions. For example, $NaHCO_3$ (baking soda) is used in baking to make bread, cookies, and cakes because it produces carbon dioxide gas when heated. You can imagine the effects of using too little or too much baking soda. To prepare a chemical reaction in a laboratory, you have to know how much of each reactant (that is, their masses) to add to obtain a certain amount of product. The study of mass relationships in chemical reactions is called **stoichiometry** (stoy key ah′ ma tree), a word coined by *Jeremias Richter* in 1792. This unfamiliar-sounding word comes from a combination of the Greek words for something that can't be divided, στοιχειον, and determining relative magnitudes, μετρειν. In your general chemistry course, you will have to solve many stoichiometric problems. Although these types of problems sometimes give begining chemistry students a lot of trouble, the logic and the procedure for solving them are really quite straightforward if you understand the basic ideas behind it.

The Masses of Atoms, Molecules, and Ionic Compounds

How much do they weigh, and what's the easiest way to weigh them?

Chemical equations tell us how many atoms, molecules, or ions react to form a certain number of atoms, molecules, or ions, but they don't tell us anything about their masses (that is, how much we have to weigh out to mix them together). It's not difficult to understand the connection between atoms and masses in chemical reactions. For example, if we wanted to prepare water from hydrogen and oxygen, we know from the balanced equation that it takes two hydrogen molecules to react with one oxygen molecule to produce two water molecules:

$$2H_2(g) + O_2(g) \longrightarrow 2H_2O(l)$$

Does this mean we can simply combine 2 g of hydrogen with 1 g of oxygen to get 2 g of water? No, that would be like preparing a ham sandwich by combining 1 lb of ham with 2 lb of bread slices. Just as the ham weighs more than the bread slices, we know the oxygen molecule is heavier than the hydrogen molecule because it has many more protons, neutrons, and electrons.

In order to know how much of each molecule to use, we have to know their masses, which requires that we know the masses of the hydrogen and oxygen atoms (that is, the **atomic mass** for each element). As you may have already noticed, the atomic masses of all the elements are given in the periodic table underneath the chemical symbol. But these are not the masses in grams. The mass of an individual hydrogen atom is 1.675×10^{-24} g, and the mass of an individual oxygen atom is 2.68×10^{-23} g. Rather than deal with numbers this small, chemists have devised a much more convenient way to describe atomic masses by inventing the **atomic mass unit (amu).** The carbon-12 atom (with six protons, six neutrons, and six electrons) was chosen as the standard and was assigned a mass of exactly 12 amu. The mass in amu of each of the other atoms was then determined relative to the carbon-12 standard, using the numbers of protons, electrons, and neutrons in other atoms relative to carbon-12. (Can you figure out how this was done?) *(Check out the website for the answer.)* For example, on this scale, the atomic mass of hydrogen is 1.008 amu and that of iron is 55.85 amu.

Two lbs of bread and 1 lb of ham don't make a ham sandwich.

Looking at the atomic mass for carbon in the periodic table, you may be puzzled by the value of 12.01 amu. Can you guess why it's not exactly 12 amu? It's because carbon has two stable isotopes: C-12 (12 amu) and C-13 (13.00335 amu), and they exist in nature in constant proportions. Anywhere in the world a sample of carbon contains 98.90% C-12 and 1.10% C-13. For example, if you had 10,000 carbon atoms, on average there would be 9890 C-12 isotopes and 110 C-13 isotopes. The average atomic mass of carbon is much closer to 12 amu because there are so many more C-12 isotopes than C-13 isotopes. The atomic masses given in the periodic table are the averages of all the naturally occurring isotopes of the elements.

Once you understand the basis of the atomic mass unit, you can readily calculate the mass of molecules, called the **molecular mass,** by the following steps: (1) write the formula of the molecule, (2) identify the elements present and count the number of atoms of each element, and (3) add the atomic masses of each element. For example, the molecular formula of ethanol is C_2H_6O. There are two C atoms, six H atoms, and one O atom, so its molecular mass in amu is

$$2(12.01 \text{ amu}) + 6(1.008 \text{ amu}) + 16.00 \text{ amu} = 46.07 \text{ amu}$$

The same procedure can be applied to calculate the mass of ionic compounds. Because ionic compounds don't exist as discrete molecules (why?), the mass would be calculated for one chemical formula unit (e.g., $MgCl_2$) or simply one formula unit. The **formula unit mass** of $MgCl_2$ is

$$24.31 \text{ amu} + 2(35.45 \text{ amu}) = 95.21 \text{ amu}$$

Moles and Molar Mass

The amu greatly simplifies dealing with the masses of atoms, molecules, and ionic compounds by using numbers that are easier to use than the very small numbers in grams. However, it's not very practical to use in the laboratory, where one usually weighs samples in grams. Based on the amu scale, we know that by proportion 12.01 g of carbon, 1.008 g of hydrogen, and 55.85 g of iron all should have the same number of atoms (why?). How many atoms would that be in each sample? You should be able to calculate the answer because the mass in grams of the hydrogen atom was just given. If we divide 1.008 g by 1.675×10^{-24} g per atom, we get 6.02×10^{23} atoms. Thus, 12.01 g of carbon and 55.85 g of iron also contain 6.02×10^{23} atoms of carbon and iron, respectively. That's a lot of atoms. If we had 6.02×10^{23} oranges, they would not only cover the entire surface of Earth, but the layers of oranges would extend over 100 miles toward outer space (imagine that!). This number is called **Avogadro's number,** after the Italian scientist ***Lorenzo Avogadro*** (actually, Lorenzo Romano Amadeo Carlo Avogadro de Quarequa e di Cerrito). His most famous work, Avogadro's law, became the basis for determining atomic masses in the late 19th century.

For doing experiments in a laboratory (where it's easier to use quantities in grams than amu), it would be more convenient to weigh elements and compounds in units of Avogadro's number just as it's more convenient to sell eggs in units of 12, or a dozen. By analogy with calling a unit of 12 a dozen, we call a unit of 6.02×10^{23} a **mole.** While the mole unit may seem intimidating because it represents such a huge number, you should convince yourself that it's just a unit representing a specific number—no different from a pair or a dozen. Whenever we have an Avogadro's number of something, we have a mole of that same thing. It's obviously not convenient to use the mole for oranges or eggs, but it is convenient for atoms because they are so small. So 1 mole of the carbon-12 isotope contains 6.02×10^{23} C-12 atoms and weighs exactly 12 g. Again, the comparison with a dozen should help you understand the mole concept:

1 mole of C-12 atoms	1 dozen brown eggs
6×10^{23} C-12 atoms	12 brown eggs
12 g	600 g (an estimate, depending on the health and size of the hens)

Because the mass of 1 mole of the carbon-12 isotope is exactly 12 g, we say that carbon-12 has a **molar mass** of 12 g. Likewise, the molar mass of hydrogen is 1.008 g and that of iron is 55.85 g. So the molar mass of an element (in grams) is numerically equal to its atomic mass, and therefore, we can get its value from the periodic table. Take a moment to look at the periodic table and check out the molar masses of some of the elements. From the molar masses of the elements, we can readily calculate the molar masses of compounds using the same procedure we saw earlier for molecular mass and formula unit mass. Referring back to the ethanol and magnesium chloride examples, we find that the molar masses of those compounds are 46.07 g and 95.21 g, respectively.

When carrying out reactions, chemists (and chemistry students) must routinely convert grams to moles and moles to grams. Of course, you will also be asked to do these conversions on homework problems and on exams. Let's do some sample exercises using the relationship between moles and grams. The conversion between mole units and gram units involves a technique that you will have to use in many types of chemistry problems. To convert one unit to another unit, you simply multiply by a factor, a conversion factor, that equates the two units such that the units you want to change are canceled and replaced by the units you want to obtain using simple multiplication. The equation would take the general form shown in the following, where n and m would be numbers that make the conversion factor equal to one:

$$\cancel{\text{unit 1}} \times \frac{n \text{ unit 2}}{m \cancel{\text{ unit 1}}} = \left(\frac{n}{m}\right) \text{unit 2}$$

As an example of a simple unit conversion, let's consider calculating the number of inches in 4 feet. Because 12 in = 1 ft, that is, the conversion factor is written as

$$\frac{12 \text{ in}}{1 \text{ ft}}$$

we would set up the equation as

$$4 \cancel{\text{ ft}} \times \frac{12 \text{ in}}{1 \cancel{\text{ ft}}} = 48 \text{ in}$$

Note that because the conversion factor, 12 in/1 ft, is equal to one, multiplication only changes the units. You can use this method to convert between any units: miles to meters, quarts to liters, and so on. For conversions between moles and grams, the conversion factor is molar mass in g/mol.

Calculate the number of grams in 5.32 moles of iron (Fe).

4.1

Answer   *The molar mass of Fe is 55.85 g, so the conversion factor is 55.85 g Fe/mol Fe. (The unit of mole is mol.) The equation can be set up as*

$$5.32 \cancel{\text{ mol Fe}} \times \frac{55.85 \text{ g Fe}}{1 \cancel{\text{ mol Fe}}} = 297 \text{ g Fe}$$

Note that we used only three figures in our answer because of the rules for significant figures. (Check out the website for a discussion of significant figures.)

EXAMPLE

You may be troubled by fractional numbers of moles if you think of the analogy between a mole and a dozen—you wouldn't normally consider 5.32 dozen because it would involve splitting eggs. But remember that we are dealing with a huge number of atoms; 5.32 mole is $5.32 \times 6.02 \times 10^{23}$ or 3.15×10^{24} atoms. You should be no more concerned with fractional numbers of moles than you would be for half-a-dozen eggs.

4.2

EXAMPLE

Calculate the number of moles in 73.96 g of calcium (Ca).

Answer   *Because the molar mass of Ca is 40.08 g, the conversion factor is 40.08 g Ca/mol Ca. When converting grams to moles, you have to invert the molar mass, but that's easy to do:*

$$73.96 \text{ g } \cancel{Ca} \times \frac{1 \text{ mol Ca}}{40.08 \text{ g } \cancel{Ca}} = 1.845 \text{ mol}$$

4.3

EXAMPLE

Calculate the mass of one gold (Au) atom in grams.

Answer   *This problem is not quite as straightforward as the gram to mole conversions, but as long as you keep the units straight, you'll be okay. The unit you want in your answer is atoms per gram. The molar mass of Au is 197.0 g, and 1 mole of Au contains 6.02×10^{23} Au atoms. Putting all this together, we have*

$$\frac{197.0 \text{ g Au}}{1 \text{ } \cancel{mol \text{ } Au}} \times \frac{1 \text{ } \cancel{mol \text{ } Au}}{6.02 \times 10^{23} \text{Au atoms}} = 3.27 \times 10^{-22} \text{ g/atom}$$

(Check out the website for more problems.)

Stoichiometry | *(στοιχειονμετρειν)*

WHY THE NAME STOICHIOMETRY? IT'S GREEK FOR MEASURING THE RELATIVE MAGNITUDES OF THINGS THAT CANNOT BE DIVIDED.

Jeremias Richter

Once you have mastered the conversions between moles and grams, you will be ready to deal with calculations involving chemical reactions (that is, stoichiometry). Let's consider the formation of water again to illustrate the procedure. When hydrogen is burned in oxygen gas or a mixture of the gases is subject to an electric spark, the gases react instantly (and often violently) to form water.

$2H_2(g)$	$+$ $O_2(g)$	$\longrightarrow$	$2H_2O(l)$
2 molecules	1 molecule		2 molecules
$2(6.02 \times 10^{23})$ molecules	6.02×10^{23} molecules		$2(6.02 \times 10^{23})$ molecules
2 mol	1 mol		2 mol

As you can see, because two molecules of H_2 react with one molecule of O_2 to produce two molecules of H_2O, the same reaction must also occur if we change the numbers to multiples of Avogadro's number or number of moles. Thus, the relative numbers of moles are identical to the relative number of molecules. This conversion allows us to conveniently study the reaction in the laboratory in terms of measurable, macroscopic quantities; that is, we use moles to get grams. Because 2 moles of H_2 always react with 1 mole of O_2 to produce 2 moles of H_2O, we can represent the relationships as

$$2 \text{ mol } H_2 \triangleq 1 \text{ mol } O_2$$

$$2 \text{ mol } H_2 \triangleq 2 \text{ mol } H_2O$$

$$1 \text{ mol } O_2 \triangleq 2 \text{ mol } H_2O$$

where the symbol $\triangleq$ means "stoichiometrically equivalent to."

This stoichiometric equivalence is analogous to the equivalence we used previously in the unit conversions and can be treated similarly. That is, the ratios of numbers of

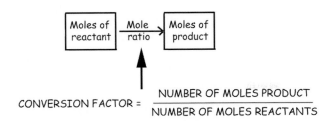

Figure 4.1 The mole ratio from the balanced equation can be used as a conversion factor to calculate moles for a reaction.

moles in the balanced equation can be used as conversion factors (Figure 4.1). All stoichiometric calculations are based on this simple mole-to-mole relationship. Let's look at an example.

4.4

EXAMPLE

How many moles of H_2O can be produced from 1.46 moles of O_2, assuming there is enough hydrogen gas present to react with all of the O_2?

Answer *According to the balanced equation, we see that 1 mole of O_2 produces 2 moles of H_2O; therefore, we can use the ratio as a conversion factor*

$$\frac{2 \ mol \ H_2O}{1 \ mol \ O_2}$$

For 1.46 moles of O_2, the number of moles of H_2O formed can be determined as

$$1.46 \ \cancel{mol \ O_2} \times \frac{2 \ mol \ H_2O}{1 \ \cancel{mol \ O_2}} = 2.92 \ mol \ H_2O$$

Analogous to the prior unit conversions, the mole ratio always takes the form such that the final quantity we want appears in the numerator. Because we usually conduct experiments by measuring the mass of substances (in grams), we can extend the previous procedure by including a conversion factor using molar mass (Figure 4.2).

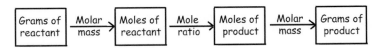

Figure 4.2 Steps for solving stoichiometric problems involving grams of reactants and products

Consider the following example.

4.5

EXAMPLE

A quantity of 0.375 g of H_2 is reacted with enough O_2 to form H_2O. How many grams of H_2O are formed?

Answer *In this problem (which is typical of many of the problems you will encounter), you must convert the number of grams of hydrogen to moles hydrogen using molar mass as a conversion factor; then moles hydrogen to moles water using the mole ratio from the balanced equation; and finally, moles water to grams water using molar mass as a conversion factor. The molar mass of H_2 is 2.016 g, and the molar mass of water is 18.02 g. It is simplest to set up the problem as a string of unit conversions. If the units cancel out such that you end up with the units you want in your answer (in this case, grams of water), you*

—Continued next page

Continued—

can be confident that the answer is right (unless you push a wrong button on your calculator). Of course, you can always do the calculation twice to be sure you haven't hit the wrong number. The number of grams of water produced is

$$0.375 \text{ g } H_2 \times \frac{1 \text{ mol } H_2}{2.016 \text{ g } H_2} \times \frac{2 \text{ mol } H_2O}{2 \text{ mol } H_2} \times \frac{18.02 \text{ g } H_2O}{1 \text{ mol } H_2O} = 3.35 \text{ g } H_2O$$

We can summarize the steps in Example 4.5 as follows:

1. Write the balanced equation for the reaction.

$$2H_2(g) + O_2(g) \longrightarrow 2H_2O(l)$$

2. Treat the coefficients as number of moles and set up mole ratios between reactants and products. The conversion factor between H_2O and H_2 is

$$\frac{2 \text{ mol } H_2O}{2 \text{ mol } H_2}$$

3. Convert the amount of the given substance in grams (which could be a reactant or a product) into the number of moles using molar mass as a conversion factor:

$$0.375 \text{ g } H_2 \times \frac{1 \text{ mol } H_2}{2.016 \text{ g } H_2} = 0.186 \text{ mol } H_2$$

4. Convert the number of moles of the given substance to the number of moles of the substance you want using the mole ratio:

$$0.186 \text{ mol } H_2 \times \frac{2 \text{ mol } H_2O}{2 \text{ mol } H_2} = 0.186 \text{ mol } H_2O$$

5. Convert the number of moles of the substance you want to grams using molar mass as a conversion factor:

$$0.186 \text{ mol } H_2O \times \frac{18.02 \text{ g } H_2O}{1 \text{ mol } H_2O} = 3.35 \text{ g } H_2O$$

As indicated in Step 3, you may be given the amount of product and asked to calculate the amounts of reactants needed to produce that amount. In fact, this is often what chemists do because they are interested in obtaining a certain amount of product. In this case, the procedure is still the same. For example, try to solve the following problem: *How many grams of O_2 are needed to produce 25.2 g of H_2O?*
The answer is 0.70 mol O_2. If you get a different answer, you can find the solution on the website along with more problems.

Limiting Reagent

We hope at this point you are convinced that doing stoichiometric problems is really quite straightforward. If you are not, you should try some more problems; we can assure you that is all it will take. The key, of course, is to convert everything into moles and then into grams or other units as needed. But the story doesn't end here because there's a practical aspect of stoichiometry that we haven't yet discussed. In real life, chemists almost never use reactants in exactly the right proportions to carry out a reaction. The reason is that the goal of a reaction is to produce the largest possible quantity of a useful compound from a given quantity of the starting material.

It is often the case that an excess of one reactant is needed to ensure that the more expensive reactant is completely converted to the desired product. Referring to Example 4.5, we would not use exactly 1 mole of O_2 to react with exactly 2 moles of H_2 to make water. One of the reactants is usually present in an amount larger than that needed to make the exact mole ratio indicated by the balanced equation. In this case, we would use a large amount of the less expensive O_2 to completely react with H_2.

There are many familiar food analogies that illustrate the principle of using an excess of something when two or more things must combine to give a final product. The difference in preparing food is that you usually want exactly the right amounts. For example, if you were preparing 10 hot dogs with buns, you would need 10 hot dogs and 10 buns. If you had 20 buns and 10 hot dogs, you still could prepare only 10 hot dogs with buns, and you'd have an excess of 10 buns. In this case, the number of hot dogs limits the number of hot dogs with buns you can prepare. It's exactly the same idea in chemical reactions, except we often want to use more of one reagent to ensure the maximum amount of product for the minimum cost as described previously.

Let's say that at the start of the reaction between H_2 and O_2, we have 5.32 moles of H_2 and 4.12 moles of O_2. How can we tell which of the two reactants is present in excess, that is, in an amount more than required to react with the other reactant? An easy way, and one that makes sense, is to ask which of the two quantities would give a smaller amount of H_2O (like we did with the hot dogs). Starting with 5.32 moles of H_2, we calculate the amount of H_2O produced as follows:

$$5.32 \; \cancel{\text{mol } H_2} \times \frac{2 \text{ mol } H_2O}{2 \; \cancel{\text{mol } H_2}} = 5.32 \text{ mol } H_2O$$

Similarly, we find the number of moles of H_2O that can be produced from 8.76 moles of O_2 to be

$$4.12 \; \cancel{\text{mol } O_2} \times \frac{2 \text{ mol } H_2O}{1 \; \cancel{\text{mol } O_2}} = 8.24 \text{ mol } H_2O$$

Thus, a smaller amount of H_2O is produced when all of H_2 is reacted. At the end of the reaction, only the product, water, and some of the original O_2 remain. For this reason, we say that O_2 is the **excess reagent** and H_2 is the **limiting reagent** because it limits the amount of the product that can be obtained. In this case, the maximum amount of water we can produce is 5.32 moles. Just like the hot dog example where having 100 buns wouldn't help us make more hot dogs with buns, it wouldn't matter if we added 10 times more oxygen—the amount of water produced is limited only by the amount of hydrogen present.

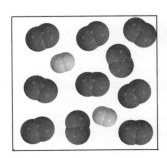

Reaction of excess oxygen with hydrogen to produce water. Hydrogen is the limiting reagent, so unreacted oxygen is left over.

We can also determine the amount of oxygen left over by subtraction (just like we did for the hot dog buns). First, we calculate the number of moles of O_2 (buns) needed to react with 5.32 moles of H_2 (hot dogs):

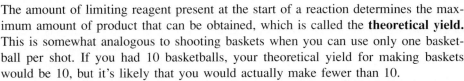

$$5.32 \; \cancel{\text{mol } H_2} \times \frac{1 \text{ mol } O_2}{2 \; \cancel{\text{mol } H_2}} = 2.76 \text{ mol } O_2$$

Then, we subtract this number from the amount of O_2 we started with. At the end of the reaction, the number of moles of O_2 left over is $(8.24 - 2.76)$ mol or 5.48 mol.

The stoichiometry story is almost complete. By determining the limiting reagent, you can calculate the maximum amount of product that could be formed, using the basic steps for solving stoichiometric problems. Remember: First convert the amounts of reactants into moles, then determine which reactant is the limiting reagent. Of course, if you are told that one of the reactants is in excess, you don't need to do this step.

Yield of a Reaction

The amount of limiting reagent present at the start of a reaction determines the maximum amount of product that can be obtained, which is called the **theoretical yield.** This is somewhat analogous to shooting baskets when you can use only one basketball per shot. If you had 10 basketballs, your theoretical yield for making baskets would be 10, but it's likely that you would actually make fewer than 10.

For chemical reactions, theoretical yield corresponds to the situation in which all the limiting reagent has reacted to give product. In practice, we almost never get the theoretical yield. The reaction may be reversible (that is, some products react to form reactants), or the products may undergo further reactions to form other products. It may also be impossible to recover all of the product from the reaction mixture. So chemists often have to settle for what is actually obtainable from a reaction. In such cases, the amount of the product obtained is called the **actual yield.** As a measure of the success of an experiment, chemists calculate the **percent yield,** which describes the proportion of the actual yield to the theoretical yield as follows:

$$\% \text{ yield} = \frac{\text{actual yield}}{\text{theoretical yield}} \times 100\%$$

Using the previous example for limiting reagent, we see that when 5.32 moles of H_2 react with 4.12 moles of O_2, 5.32 moles of H_2O are formed. This is the theoretical yield. If only 4.88 moles of H_2O are isolated at the end, then the percent yield is given by

$$\% \text{ yield} = \frac{4.88 \text{ mol}}{5.32 \text{ mol}} \times 100\% = 91.7\%$$

It's Time to Practice!

If you don't feel like you completely understand all of the concepts discussed so far in this chapter, you should go back and read it again. Once you feel comfortable with it all, it's time to practice because, like anything else from playing the piano to playing basketball, understanding how to do something isn't enough to master it. To master solving stoichiometric problems, you have to practice solving them (and there are lots of them on the website). The good news is that it actually takes less time to master solving stoichiometric problems than you think. Work on enough problems to feel confident in your ability to solve them. At first you might want to have the book handy as a reference, but realize that you won't have it to use on tests.

More Stoichiometry for the Daring

Caution: Do not read this until you have mastered the previous material.

We have covered everything you need to know about stoichiometry when reactions involve substances that you can weigh, which includes just about anything if you have the right equipment. But, often it's more convenient to measure a chemical reagent that is not a solid using its volume (if it's a liquid, gas, or solution) or its pressure (if it's a gas). The basic steps in solving stoichiometric problems still apply, but you need to use different conversion factors for determining the number of moles. To gain further understanding of stoichiometry problems, let's consider cases that involve liquids, gases, and solutions where the mass is not the measured quantity.

Liquids

You may encounter problems in which the volume of a pure liquid is given instead of its mass. In such a case, you would also be given the density of the liquid. Chemists often have to deal with measuring liquids in chemical reactions, but they have to look up the density in a handbook (unless it's given on the reagent bottle). The density of a liquid is given in units of grams per milliliter (g/mL). Because density tells us the number of grams in a milliliter, it can be used as a conversion factor, like 12 in/ft, because mL $\times$ g/mL gives the number grams. So if we know the density and the volume, we can easily calculate the mass, and once we have the mass, the problem becomes just like the problems we have already discussed, as shown in Figure 4.3.

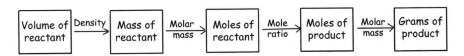

Figure 4.3 Steps for solving stoichiometric problems involving volumes of liquid reactants

Consider the following example.

EXAMPLE 4.6

How many grams of carbon dioxide gas are produced by the complete combustion of 1.0 L of octane, C_8H_{18}, a typical component of gasoline that has a density of 0.70 g/mL?

$$2C_8H_{18}\,(l) + 25O_2(g) \longrightarrow 18H_2O(g) + 16CO_2(g)$$

Answer *Because we are told that the combustion is complete, we assume that oxygen is in excess and we don't have to determine the limiting reagent. The steps in solving the problem involve converting liters of octane to grams octane using its density as a conversion factor; then converting grams of octane to moles octane using its molar mass as a conversion factor; then converting moles octane to moles carbon dioxide using the mole ratio from the balanced equation; and finally, converting moles carbon dioxide to grams using its molar mass:*

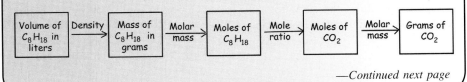

—Continued next page

Continued—

We could add the conversion of liters to milliliters in the mix, given that 1 liter (L) =1000 mL, but it's easier to just start with 1000 mL instead of 1.0 liter. After checking to see that all the units cancel except g CO_2, we obtain the desired result as shown in the following:

$$1000 \text{ mL octane} \times \frac{0.70 \text{ g octane}}{1 \text{ mL octane}} \times \frac{1 \text{ mol octane}}{114 \text{ g octane}} \times \frac{16 \text{ mol } CO_2}{1 \text{ mol octane}}$$

$$\times \frac{44.0 \text{ g } CO_2}{1 \text{ mol } CO_2} = 4.3 \times 10^3 \text{ g } CO_2$$

Solutions

Chemists often carry out chemical reactions involving prepared aqueous solutions containing known amounts of dissolved reagents (the solute). A familiar example of such a solution is vinegar; on average, there are 5 g of acetic acid in every 100 mL solution of water. You could easily convert a known volume of vinegar to grams acetic acid by multiplying the number of milliliters times 5g/100 mL, just as we did with density in Example 4.6. For this solution, we say that 5g/100mL is the **concentration** of acetic acid. Concentration simply refers to the amount of a substance present in a given amount of solution. It would be easier to do stoichiometric calculations if the concentrations were in moles instead of grams. For this reason, chemists use a concentration term called **molarity,** which is the number of moles of a substance in a liter of solution, or mol/L. Concentrations expressed in mol/L are called *molar concentrations.* They make stoichiometric calculations relatively easy because molarity can be used as a conversion factor between volume (L) and moles. The abbreviation for molarity is given as *M,* which means mol/L. (*Check out the website for more discussion of molarity and for problems on molarity calculations.*)

Figure 4.4 shows the steps for solving stoichiometry problems using molarity. This procedure is illustrated in Example 4.7.

Figure 4.4 Steps for solving stoichiometric problems involving volumes of solutions with known molarity

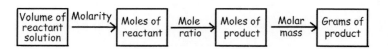

4.7

EXAMPLE

Oxalic acid ($H_2C_2O_4$) is used for the removal of rust (Fe_2O_3). Calculate the number of grams of rust that can be removed with 1.00 L of 0.100 M solution of oxalic acid. The balanced equation is shown as follows.

$$Fe_2O_3(s) + 6H_2C_2O_4(aq) \longrightarrow 2Fe(C_2O_4)_3^{3-}(aq) + 3H_2O(l) + 6H^+(aq)$$

Answer *The equation tells us that it takes 6 moles of $H_2C_2O_4$ to consume 1 mole of Fe_2O_3. From the definition of molarity, we see that there is 0.100 mole of $H_2C_2O_4$ in 1.00 L of the solution. Knowing the molar mass of Fe_2O_3, we can set up the following chain of conversion factors:*

$$1.00 \; L \; H_2C_2O_4 \; soln \times \frac{0.100 \; mol \; H_2C_2O_4}{L \; H_2C_2O_4 \; soln} \times \frac{1 \; mol \; Fe_2O_3}{6 \; mol \; H_2C_2O_4}$$

$$\times \frac{159.7 \; g \; Fe_2O_3}{1 \; mol \; Fe_2O_3} = 2.66 \; g \; Fe_2O_3$$

Gases

We've already seen some reactions involving gases, and, as you might imagine, it's not always easy to weigh them. What can you measure to determine the number of moles of a gas? To begin with, you should realize that gases, unlike liquids and solids, do not possess a constant volume (V); their volume is determined by that of the container. Therefore, a large volume of gas does not necessarily mean a lot of gas. The amount of the gas present also depends on its temperature (T) and pressure (P). There is a very useful equation that relates V, T, P, and the number of moles (n) for a so-called ideal gas. It is appropriately called the **ideal gas equation:** $PV = nRT$, where R is the gas constant. In stoichiometric problems, we usually assume the gas is ideal, and you can use the equation even if you don't know what an ideal gas is. (*Check out the website for a discussion of ideal gases and the ideal gas equation.*) So if we are given P, V, and T of a gas, we can rearrange the equation to get $n = PV/RT$. In calculating n, P must be in atm (atmospheres), V in L (liters), T in K (kelvins), and R is given by 0.0821 L $\cdot$ atm/K $\cdot$ mol. Remember to convert temperature in Celsius to kelvins by adding 273. (*Check out the website for a discussion of temperature scales.*)

Figure 4.5 shows the steps for solving stoichiometric problems involving gases.

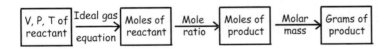

Figure 4.5 Steps for solving stoichiometric problems involving gases of known volume, pressure, and temperature

4.8

EXAMPLE

Lithium hydride (LiH) reacts with water to form hydrogen gas according to the equation

$$LiH(s) + H_2O(l) \longrightarrow LiOH(aq) + H_2(g)$$

During World War II, U.S. pilots carried LiH tablets to fill their life belts and lifeboats with hydrogen gas in the event of a crash in the ocean. How many grams of LiH are needed to fill a 4.10-L life belt at 0.970 atm and 12°C?

—Continued next page

Continued—

Answer *In this problem, we're given the product and asked to determine the amount of reactant, but the general procedure is still the same. Of course, we assume there is an excess of water in the ocean, so we don't need to worry about which reactant is the limiting reagent. First, we must determine the number of moles of H_2 using the ideal gas equation.*

$$n = \frac{PV}{RT}$$

$$= \frac{(0.970\ atm)(4.10\ L)}{(0.0821\ L \cdot atm/K \cdot mol)(12 + 273)K} = 0.170\ mol$$

Now continue with the usual steps to get the amount of LiH in grams:

$$0.170\ \cancel{mol\ H_2} \times \frac{1\ \cancel{mol\ LiH}}{1\ \cancel{mol\ H_2}} \times \frac{7.95\ g\ LiH}{1\ \cancel{mol\ LiH}} = 1.35\ g\ LiH$$

It's time to practice again, so check out the website for a wide variety of stoichiometric problems. Figure 4.6 summarizes the frequently encountered cases in stoichiometric problems, but it might be helpful for you to try to construct a similar chart without looking at this one.

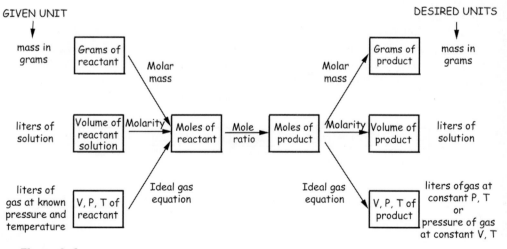

Figure 4.6   Steps for solving stoichiometric problems involving solids, solutions, or gases as reactants or products.   First, identify the units you are given to determine which conversion factor to use to get moles of reactant. Then, convert to moles product. Finally, identify the units you want in your answer and choose the conversion factor that will convert moles product to desired units.

Test your understanding of the material in this chapter

IMPORTANT TERMS

Explain the following terms in your own words:

actual yield, p. 54
atomic mass, p. 47
atomic mass unit, p. 47
Avogadro's number, p. 48
concentration, p. 56
excess reagent, p. 53

formula unit mass, p. 47
ideal gas equation, p. 57
limiting reagent, p. 53
molarity, p. 56
molar mass, p. 48

mole, p. 48
molecular mass, p. 47
percent yield, p. 54
stoichiometry, p. 46
theoretical yield, p. 54

CONCEPTS

Explain in your own words:

- The atomic mass unit, molecular mass, the mole, molar mass, and how Avogadro's number is used in calculations.
- The mole method in stoichiometric calculations.
- The relationships among actual yield, theoretical yield, and percent yield.
- How to calculate the number of moles using the ideal gas equation and using the volume and molarity of a solution.

UNDERSTANDING CHEMISTRY

To test your understanding of the material in this chapter, let's consider the following case of preparing an important industrial chemical. Fertilizers are essential for producing large and healthy crops to feed the world's rapidly increasing population. Among the common fertilizers is urea [$(NH_2)_2CO$], which can be prepared by bubbling carbon dioxide (CO_2) gas into an aqueous ammonia (NH_3) solution:

$$CO_2(g) + 2NH_3(aq) \longrightarrow (NH_2)_2CO(aq) + H_2O(l)$$

Industrially, millions of tons of urea are manufactured every year. Here we will consider the reaction on a much smaller scale carried out in the laboratory. Suppose you are provided with the reactants in the following manner: CO_2: 24.2 liters of the gas measured at 25°C and 3.60 atm; NH_3: 1.84 liters of a 0.376 M NH_3 solution. How many grams of urea will form from these two substances?

How should you proceed? In this chapter you learned the mole method for solving stoichiometric problems, so the first step is to convert the reactants into number of moles. For CO_2, we use the ideal gas equation. After converting 25°C to 298 K, we write

$$n = \frac{PV}{RT}$$

$$= \frac{(3.60 \text{ atm})(24.2 \text{ L})}{(0.0821 \text{ L} \cdot \text{atm/K} \cdot \text{mol})(298 \text{ K})}$$

$$= 3.56 \text{ mol}$$

For NH_3, we calculate the number of moles like this:

$$\text{no. moles of } NH_3 = 1.84 \text{ L} \times \frac{0.376 \text{ mol}}{1 \text{ L soln}}$$

$$= 0.692 \text{ mol}$$

The next step is to determine which of the two reactants is the limiting reagent. From the balanced equation, we see that 1 mole of CO_2 produces 1 mole of urea, so 3.56 moles of CO_2 will yield 3.56 moles of urea. On the other hand, 2 moles of NH_3 produce 1 mole of urea, so 0.692 mole of NH_3 will result in $\frac{0.692}{2}$ or 0.346 mole of urea. Therefore, NH_3 must be the limiting reagent and CO_2 the excess reagent. (Given the fact that ammonia is about 10 times more expensive than CO_2, is the result surprising?)

Assuming all the limiting reagent is used up in the reaction, we will obtain 0.346 mole of urea, and from the molar mass of the compound (60.06 g/mol), we calculate the number of grams as follows:

$$0.346 \text{ mol } (NH_2)_2CO \times \frac{60.06 \text{ g } (NH_2)_2CO}{1 \text{ mol } (NH_2)_2CO} = 20.8 \text{ g } (NH_2)_2CO$$

This quantity, as you recognize, is the theoretical yield of the reaction. Suppose in practice the amount obtained is 18.7 g, which is the actual yield. The percent yield, then, is given by

$$\% \text{ yield} = \frac{18.7 \text{ g}}{20.8 \text{ g}} \times 100\%$$

$$= 89.9\%$$

Industrial chemists spend a great deal of effort to enhance the percent yield of a reaction. Because huge quantities of materials are involved, even a slight improvement of the reaction by 1% in the yield, say, can result in a substantial saving in cost.

The Energy of Chemistry
Some like it hot

When you turn on the ignition key in your car, you start a reaction between gasoline and oxygen to produce carbon dioxide and water and a lot of **energy.** In Chapters 3 and 4, you learned that atoms, molecules, and ions recombine to form different products, and you know how to determine the type and amount of product we can get from a reaction. If you wanted to calculate the amount of carbon dioxide produced by the combustion of a gallon of gasoline, you could do it if you had the appropriate conversion factors. In this chapter we will consider the *energy* part of the equation. In the reaction between gasoline and oxygen, energy is produced because the product molecules formed are more stable than the reactant molecules. If this seems hard to imagine, consider a boulder sitting precariously on a cliff that falls to the ground with a crash. As it reaches its final, stable state (lower gravitational potential energy), it generates a lot of energy, which we recognize by the crashing sound and the broken pieces of rock. Similarly, energy changes in chemical reactions may be perceived as an explosive bang, or if the energy change is small, you may only observe a slight change in temperature. Some chemical reactions are more important for the energy they produce than for the products that are formed. They provide a source of energy for heating, cooking, transportation, industrial processes, and a variety of recreations.

Energy and Heat

Energy is a much-used term that represents a very abstract concept. If you feel tired, you might say you haven't any energy, or you might eat a candy bar in order to get some extra energy before exercising. On the other hand, people who care about the environment say we need to find alternatives to nonrenewable energy sources such as oil and coal. Is this the same kind of energy you get from a candy bar?

The energy released from some chemical reactions is explosive. Always wear eye protection when working in a chemistry laboratory.

What exactly is energy? It's not easy to define energy, but we know it when we experience it. Unlike matter, energy is known and recognized by its effects, not by any perceptible characteristic; it can't be seen, touched, smelled, or weighed. A useful way to define energy is to describe its many different forms. There are three that are of particular interest to chemists: radiant, thermal, and chemical energy.

Radiant Energy

Microwaves, radio waves, gamma rays, X rays, light—this is the energy that travels through space. We'll discuss this in more detail in Chapter 6. The light emitted from a hot object, like a piece of hot tungsten filament wire in a light bulb, is radiant energy. The energy from the sun (a very hot object), or solar energy, is another example of radiant energy. Solar energy heats the atmosphere and Earth's surface, stimulates the growth of vegetation through the process called photosynthesis, and influences global climate patterns.

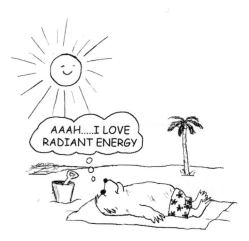

Thermal Energy

Molecules in motion—this is the energy associated with the random motion of atoms and molecules. Imagine a water molecule in the gaseous state. It can move through space by linear, direct movements at a certain speed; this is called translational motion. It can also move internally because it's not rigid. The H and O atoms rotate around the chemical bonds holding them together like wheels on an axle. The molecule also vibrates as if the bonds holding the atoms together were stiff springs. **The greater the extent of molecular motion (translational, rotational, and vibrational) in a sample of matter, the greater is the thermal energy and the higher the temperature of the matter.**

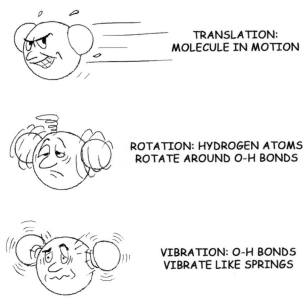

TRANSLATION:
MOLECULE IN MOTION

ROTATION: HYDROGEN ATOMS
ROTATE AROUND O-H BONDS

VIBRATION: O-H BONDS
VIBRATE LIKE SPRINGS

The thermal energy of water molecules
is the sum of its translational,
rotational, and vibrational energies.

Heat, another much used term for a really abstract concept, is the transfer of thermal energy between two bodies at different temperatures. For example, if we put an ice cube in a glass of hot water, some of the thermal energy of the "hot" water molecules is transferred to the "cold" water molecules when fast-moving molecules collide with slow molecules to give them some extra energy and break loose from the rigid ice structure. Eventually, the average energy of the water molecules is somewhere between that of the ice and the hot water before they were combined.

Chemical Energy

Bond making—this is the energy released when a bond is formed between atoms or ions (that is, when they stick together). In Chapter 7 we will discuss chemical bonds in more detail, but we want to say a little about bonding now so you can better understand chemical energy. There are different types of forces (or bonds) that hold atoms, molecules, and ions together in compounds. One of the strongest of these forces is the chemical bonds (called **covalent bonds**) that hold atoms together by sharing electrons—like the bonds between H and O in water. Because these are very strong forces, the formation of these bonds releases a large amount of energy. While it may be hard to imagine how forming bonds releases energy, it's easier to understand the energy associated with bond formation if you consider the reverse process—breaking a bond. To separate the atoms held together by a chemical bond requires the input of energy. It's like trying to pull apart two balls that are glued together. You know from experience that it takes a certain amount of energy to do this. Conversely, when atoms

It takes a lot of energy to break
a bond between two atoms.

come together to form a bond, an equal amount of energy is released. This is an important point about energy changes: **If we reverse a process, the magnitude of the energy change is the same, but the effect is just the opposite.** That is, if a process absorbs energy in one direction, an equal amount of energy will be released if the process goes in the other direction.

Another very strong force that results in the release of chemical energy is the attraction between two oppositely charged ions (called an **ionic bond**). It requires an input of energy to separate oppositely charged ions just as it would to separate the opposite poles of two magnets. According to the rule, an equal amount of energy is released when they come together.

The weakest force that holds atoms and molecules together is that which typically occurs between molecules in the liquid or solid state, like the forces that hold water molecules together in ice. These are called **intermolecular forces** (IMF) because they exist between molecules. These are the forces that must be broken when a substance goes from liquid to gas (that is, when it evaporates). For all of these chemical forces, the amount of energy required to overcome them is determined by the type and arrangement of constituent atoms, molecules, or ions. When substances participate in chemical reactions (that is, breaking and making bonds), chemical energy is absorbed, released, or converted to other forms of energy. Similarly, when substances undergo phase changes (like freezing or evaporating), energy is absorbed, released, or converted to other forms of energy because intermolecular forces are formed or broken.

There is another very important force involved in many chemical reactions, but it occurs within atoms, not between them. It is the force of attraction between the nucleus of an atom and its electrons. Recall that a major type of chemical reaction is oxidation-reduction, or the transfer of electrons from one atom to another. As you might imagine, it requires a lot of energy to remove an electron from an atom. This energy is called *ionization energy,* and we will say more about it in Chapter 6.

Table 5.1 lists the range of energies for different types of chemical forces.

Table 5.1 Range of Energies for Chemical Forces

Chemical Force	Energy (kJ/mol)
Covalent bond	200–800
Ionic bond	600–4000
IMFs	10–60
Ionization energy	400–2400

Energy Is Conserved | *It's a law!*

The **law of conservation of energy** states that energy can be converted from one form to another, but it can't be created or destroyed. In other words, the energy of the universe is always the same. This is why the energy change when we reverse a process is equal in magnitude but opposite in sign. If you eat a candy bar before jogging, the enzymes[1] in your body break down the molecules in the candy in a process called *metabolism.* The energy released is mostly converted to the kinetic energy associated with jogging. The law of conservation of energy says that the amount of energy given off by the digestion of the candy bar must be equal to that converted to kinetic energy as well as the energy that is simply dissipated as heat. The law of energy conservation is extremely important because it provides the basis for measuring energy changes, an essential part of chemistry. Measuring energy changes in chemical reactions, called **thermochemistry,** is important for understanding the nature of physical, chemical, and biological processes; it is also of practical value if we are interested in the industrial preparation of chemicals or in looking for new energy sources.

Before we talk about measuring energy changes, let's consider a commonplace example of an energy change. Imagine boiling water on a gas stove. The water molecules absorb heat energy from the stove, and they start moving faster. Eventually, they move fast enough to overcome the forces (in this case, intermolecular forces) holding them together in the liquid state, and they escape into the air space. The amount of energy absorbed by the water to break intermolecular forces and increase its thermal energy is equal to the amount of energy that the flame imparted to the water. Where does the flame's energy come from? It comes from the combustion reaction of the gas (mostly methane) with oxygen to produce carbon dioxide and water and a lot of energy because the products, carbon dioxide and water, have stronger bonds and are more stable than the reactants, methane and oxygen.

If you boil water on an open stove, not all of the energy from the combustion of methane is absorbed by the water. Some of the energy is absorbed by the pot, and some radiates away from the pot and increases the thermal energy of the surrounding air molecules. What if we were interested only in the energy absorbed by the water? Chemists are faced with a similar question whenever they want to measure an energy change. To keep track of what they are measuring, chemists define that part of the universe that's of interest to them as the *system* and the rest of the universe as the *surroundings.* It's an arbitrary designation, but a very useful one because it forms the basis for calculating energy changes. For the pot of water on the stove, we could say that the system is the water, or it could be the water and the pot, or we could say that the entire kitchen, including the water, the pot, and the stove, is the system. The

---

1. Enzymes are biological catalysts that increase the speed of a reaction.

system, then, can be anything we are interested in studying. In a laboratory, the system could be a beaker of water or a mixture of oxygen and hydrogen gases in a flask. In the previous candy bar example, the system could be a human body. In any process, the system may gain or lose energy, and, because the total energy of the universe remains unchanged, the surroundings must lose or gain the same amount of energy.

To help you better understand the system/surroundings designation, let's look more closely at the two processes involved in boiling water on a stove: the evaporation of water and the burning of natural gas. The evaporation of water is an example of a physical change, which doesn't involve changing the identity of the water molecules. It is simply the separation of water molecules from one state to another according to the equation

$$H_2O(l) \longrightarrow H_2O(g)$$

Because energy is absorbed by the system (the water) from the surroundings (the flame), we call this an **endothermic process** (*endo-* is a prefix meaning "within"), where heat goes into the system. As a result, the energy content of the gaseous water molecules (also part of the system) is greater than that of the water molecules in the liquid before heating. You may be interested to know that we use the endothermic evaporation of water to cool off when we are overheated; that is, we perspire. The water in your sweat absorbs heat from the surroundings in order to evaporate. In this case, we can consider the sweat as the system. Your body, as part of the surroundings, supplies the heat and thus gets cooler.

Similarly, the melting of an ice cube is a physical process that absorbs heat. When heat is supplied to an ice cube, it begins to melt:

$$H_2O(s) \longrightarrow H_2O(l)$$

Again, this happens because energy is supplied from the surroundings to the system (the ice cube). The source of energy (the surroundings) can be your hand, a burning match, the sun, or collisions with hotter air molecules. After melting, the liquid water, which has the same mass as the original ice cube, has a higher energy content. The energy taken up by the ice cube is used to break up the three-dimensional network of water molecules, resulting in an increase in the thermal energy of the liquid water molecules.

As we mentioned earlier in the chapter, if we reverse a process, the magnitude of energy change would be the same, but the effect would be just the opposite. Therefore, if we freeze water at 0°C, energy in the form of heat would be released by the system (water) to the surroundings. Here, the process is said to be **exothermic** (*exo-* is a prefix meaning "outside"). Likewise, condensing steam to water would generate energy, and it is also an exothermic process.

The burning of natural gas to boil the water in the pot is an example of a chemical change—or chemical property. Methane (CH_4), the major component of natural gas, is produced by the bacterial decomposition of vegetable matter under water. It is also most likely the gas you use for the Bunsen burner in the laboratory. The equation for the combustion of methane in air is

$$CH_4(g) + 2O_2(g) \longrightarrow 2H_2O(l) + CO_2(g)$$

This reaction (the system in this case) gives off a lot of heat and some light (because the products are so hot they emit radiant energy in the form of a flame). Where does the energy come from? Originally, there are carbon-to-hydrogen and oxygen-to-oxygen bonds in CH_4 and O_2. At the end, we have oxygen-to-hydrogen and carbon-to-oxygen bonds in H_2O and CO_2. As mentioned earlier, to break a chemical bond, we must supply energy to the molecule, so bond-breaking processes are *always* endothermic.

Correspondingly, energy must be given off when a bond is formed, and bond-forming processes are *always* exothermic. In this particular reaction, the energy released due to bond formation in the products outweighs the energy absorbed for bond breakage in the reactants, so the net effect is a release of heat—an exothermic reaction. The use of "outweigh" is not strictly correct because energy doesn't have any mass that we can actually weigh. But thinking about energy changes in terms of a balance between two processes—the amount needed to break bonds and the amount released when bonds are formed—is a useful analogy.

You may be wondering why we have to supply energy in the form of a burning match to start the reaction. Many reactions require an activation energy to initiate the process (e.g., breaking the bonds in heated molecules). But once the first molecules react, the energy released provides the activation energy needed to drive the reaction of the remaining molecules (and the release of a lot of heat energy), and the reaction becomes self-sustaining.

An example of an endothermic reaction is the decomposition of mercury(II) oxide, HgO. At room temperature, HgO is a stable compound. When heated above 100°C, it begins to decompose into mercury and oxygen gas:

$$2HgO(s) \longrightarrow 2Hg(l) + O_2(g)$$

Because energy is absorbed by HgO (the system in this case), this is an endothermic process. Of the many reactions you will study in chemistry courses, almost all of them are either exothermic or endothermic. (A few special ones produce no heat effects.)

We have now laid the groundwork for describing the measurement of energy changes in chemical reactions. But first we should say a few words about energy units. For many years, calorie (cal) was the unit for energy. Now, by international convention, chemists have chosen joules (J) to represent energy. The relation between these two units is

$$1 \text{ cal} = 4.184 \text{ J}$$

In most cases, the energy changes will be expressed as kilojoules (kJ), where 1 kJ = 1000 J. The calories we talk about in food are really kilocalories (where 1 kcal = 1000 cal). So if a candy bar is said to "contain" 200 calories, it means that the sugar and fat molecules in the candy bar produce 200 kcal of energy when they react with oxygen to form carbon dioxide and water—the ultimate by-products of the biochemical breakdown (metabolism) in your body.

Measuring Energy Changes

Because heat changes in chemical processes tell us something about the relative stabilities of compounds, the measurement of such changes has many practical applications (for example, the cold packs and hot packs used by athletes). How might you measure the change in heat in a chemical reaction? For a lot of reactions, you can simply use a thermometer. Consider the acid-base reaction between a HCl solution and a NaOH solution. If you mix the two solutions together and place a thermometer in the mixture, you would see the temperature rise because this reaction is exothermic. To accurately quantify the heat change we would want to be sure the thermometer was measuring all of the heat released. So we need to ensure that any part of this heat doesn't escape to the surroundings. We can minimize the heat loss to the surroundings if we conduct the experiment in an insulated container. We call this container a **calorimeter,** and the experimental measurement of heat changes in chemical reactions and physical changes is called **calorimetry.** In practice, we place the reacting system in a calorimeter and monitor the temperature rise or fall (depending on whether the reaction is exothermic

Figure 5.1   A constant-pressure calorimeter

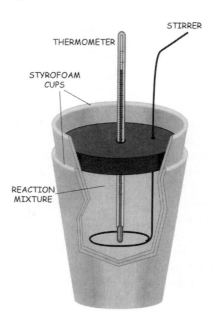

or endothermic). A styrofoam cup with a lid actually works pretty well as a simple calorimeter (Figure 5.1). Suppose we want to measure the energy change when 50 mL of a 1.0 *M* HCl solution at 20.0°C reacts with 50 mL of a 1.0 *M* NaOH solution also at 20.0°C. The chemical equation that describes the reaction is

$$HCl(aq) + NaOH(aq) \longrightarrow NaCl(aq) + H_2O(l)$$

We mix the two solutions in the calorimeter with proper stirring and note the final maximum temperature reading to be 26.7°C. So the temperature rise is 6.7°C.

    That's essentially all there is to the experiment, but how can we calculate the energy change of this acid-base neutralization reaction using the temperature change? Let's consider exactly what we are measuring. From the temperature change of the solution after mixing, we can determine the amount of heat released by the chemical reaction as it is absorbed by the molecules and ions (what are they?) in the solution. When the molecules and ions absorb the heat, their thermal energy increases and causes an increase in temperature. Different molecules and ions require different amounts of energy to increase their thermal energy. This behavior can be quantified with a property called **specific heat** (s), which is the amount of heat required to raise the temperature of 1 g of a substance by 1°C. Of course, the amount of the substance present is also important. We know from experience a certain amount of heat is needed to bring a given quantity of water to its boiling point, and it makes sense that it will always require the same amount of heat for the same amount of water to reach 100°C. On the other hand, every substance has its own specific heat. The specific heat (s) of water is 4.184 J/°C · g, meaning that it takes 4.184 J of energy to raise the temperature of 1 g of water by 1°C. (Note from the conversion factor on p. 67 that this is the same as 1 cal/°C · g.) It isn't a coincidence that the specific heat of water is exactly 1 cal/g°C. The calorie was originally defined as the amount of heat needed to raise the temperature of 1 gram of water by 1°C.

    The specific heat enables you to calculate the amount of heat (*q*) absorbed or released in a calorimetric experiement if you know the mass (*m*) and the temperature change. The change in heat is given by the equation

$$q = ms\Delta t$$

where Δt (delta t) is the change in temperature, that is

$$\Delta t = t_{\text{final}} - t_{\text{initial}}$$

The unit for q can be derived from the equation

$$q = ms\Delta t$$

$$= g \times \frac{J}{°C \cdot g} \times °C = J$$

To calculate the energy change for the neutralization reaction, we can use the specific heat of water as an approximation, because the final solution is mostly water. The mixture actually contains water molecules and Na^+ and Cl^- ions, but the number of water molecules is much greater than that of the ions. Therefore, the heat absorbed by 100 mL solution or 100 g (assuming that the solution is essentially pure water with a density of 1.0 g/mL) for the acid-base neutralization is

$$q = 100 \text{ g} \times \frac{4.184 \text{ J}}{°C \cdot g} \times (26.7 - 20.0)°C = 2800 \text{ J}$$

If all of the heat from the reaction were absorbed by the solution, this would be the value for the energy change for the reaction.

Is all the heat released by the reaction absorbed by the solution? No, some of the heat will also be absorbed by the calorimeter itself. Because we are using an insulated container, we assume that the heat absorbed by the calorimeter is negligible and none is dissipated to the surroundings. The use of a styrofoam container makes the assumption reasonably valid.

How can we distinguish between the heat absorbed by the water and the heat released by the reaction? According to the law of energy conservation, both are 1880 J, but in one case, it is the heat lost, and in the other, it is the heat gained. A simple way to distinguish the two is to assign opposite signs. By convention, we use a negative sign to indicate heat released or exothermic processes and a positive sign to indicate heat absorbed or endothermic processes.

Now that you understand the quantitative basis of calorimetry, the formal calculation of the heat released in a calorimetric experiment should seem straightforward. In this isolated (insulated) system, we know that the total heat change must be zero because no heat is transferred from the system to the surroundings. Therefore, the amount of heat lost by the reaction (a negative amount because the reaction is exothermic) plus the amount of heat gained by the solution (a positive amount because this is an endothermic process) must be equal to zero. Or the heat gained must be equal to the heat lost. Let the heat generated from the reaction be q_{rxn}, where rxn denotes reaction. Because we assume that all of the heat from the reaction is absorbed by the water, the total energy change is given as the sum of the two distinct heat values. According to the law of conservation of energy, their sum must be zero, that is

$$q_{\text{rxn}} + q_{\text{water}} = 0$$

or

$$q_{\text{rxn}} = -q_{\text{water}}$$

where q_{water} is the heat absorbed by the water. Therefore,

$$q_{\text{rxn}} = -2800 \text{ J}$$

Would the change in q_{rxn} be the same value if we had used double the amounts of solution? No, the heat of the reaction would also be doubled because the mass is twice that of the original solution. So it would be best to indicate the heat of the reactions in terms of the amount of substance. Chemists prefer to express such quantities on a per mole basis. In our example, we had 50.0 mL of 1.00 M solutions of HCl and NaOH, so there was 0.0500 L × 1.00 mol/L or 0.0500 mole each of HCl and NaOH reacting. For 1 mole of the acid and the base, then, the heat evolved would be

$$-\frac{2800 \text{ J}}{0.0500 \text{ mol}} = -56{,}000 \text{ J/mol}$$

$$= -56.0 \text{ kJ/mol}$$

This quantity is the molar heat of neutralization.

Remember to pay attention to the important sign conventions in thermochemistry: For exothermic reactions, such as the acid-base neutralization and combustion reactions discussed here, the energy change has a negative sign. Conversely, for endothermic reactions, the energy change has a positive sign. (*Check out the website for more problems on calorimetry.*)

Enthalpy | *A funny name for energy*

The amount of heat released or absorbed in a chemical process can be slightly different depending on whether the reaction occurs under conditions where the pressure is constant or under conditions where the volume is constant. The calorimeter described earlier is called a constant-pressure calorimeter because the reaction is carried out under constant atmospheric pressure conditions. Another common type of calorimeter is the constant-volume calorimeter, and as the name implies, the reaction is carried out under constant-volume conditions. This requires a strong, completely sealed container where the reaction can occur in a fixed volume.

Consider conducting a calorimetry experiment on a reaction that produces a net increase in the number of gas molecules. If this is done in a sealed chamber of fixed volume, the pressure will increase inside the chamber with the production of gas. If the same reaction is conducted in a constant-pressure calorimeter (where the calorimeter is not sealed from the atmosphere), some of the gas produced will leave the calorimeter. In this case, the amount of heat energy measured will be a bit smaller than it would be if the reaction were carried out in a constant-volume apparatus. The reason is that some of the energy is used by the molecules to push the atmosphere back in order to enter the surroundings.

In fact, the heat change measured in a constant-volume calorimeter is the total energy change of the system—all the thermal and chemical energy of the molecules, ions, and atoms. The concept of energy change is best understood if you think about it in these molecular terms. That is, the total heat change calculated from a calorimetry experiment is simply the sum of the changes in the motions of atoms, molecules, and ions and the chemical forces that glue them together. The problem is that the majority of the reactions studied by chemists are carried out with the reacting systems exposed to the atmosphere, where the heat change may not always be the same as the total energy change in the system. To deal with the problem, chemists introduced a special term for the heat change at constant pressure. The term is **enthalpy** (H) and the change in enthalpy (ΔH) is equal to the heat change at constant pressure. That is, $\Delta H = q_p$, where the subscript p denotes that this is a constant pressure process and ΔH denotes the change in enthalpy.

Although the name *enthalpy* may sound unfamiliar to you, it is practically the same as energy for our purpose. Of course, if a reaction does not involve a net change in the number of moles of gases [for example, $H_2(g) + Cl_2(g) \longrightarrow 2HCl(g)$], enthalpy change is the same as energy change. Enthalpy is often thought of as the heat content of a substance, but we can never actually determine the absolute amount of enthalpy of a substance. In practice, we can measure only the enthalpy change in a chemical process, as shown in the constant-pressure calorimetry example.

Enthalpy of Chemical Processes

The q_{rxn} we determine from constant-pressure calorimetry experiments, like the one described earlier in the chapter, is actually ΔH_{rxn}. Most energy changes associated with chemical processes are reported as enthalpy changes, so we need to think, talk, and write about the energetics of chemical reactions in terms of enthalpy changes. Let's consider the general equation

$$\text{reactants} \longrightarrow \text{products}$$

We can represent the enthalpy change of the reaction as

$$\Delta H_{rxn} = H(\text{products}) - H(\text{reactants})$$

If we knew the enthalpy values of the reactants and products, we would be able to calculate ΔH_{rxn} and, hence, the heat change without doing an experiment. But this is not possible because the absolute enthalpy of any substance is not known. However, this restriction does not prevent us from determining ΔH_{rxn}, which is equal to the heat change measured under constant-pressure conditions. For our earlier example of burning methane in air,

$$CH_4(g) + 2O_2(g) \longrightarrow 2H_2O(l) + CO_2(g)$$

calorimetric measurements show that when 1 mole of methane is burned, the heat generated is 890.4 kJ. Because this is an exothermic reaction, ΔH_{rxn} has a negative sign, so we can write $\Delta H_{rxn} = -890.4$ kJ. We can attach this ΔH_{rxn} value to the chemical equation and get a single equation (sometimes called *a thermochemical equation*) that shows reactants, products, and enthalpy change.

$$CH_4(g) + 2O_2(g) \longrightarrow 2H_2O(l) + CO_2(g) \qquad \Delta H_{rxn} = -890.4 \text{ kJ}$$

Remember that it's important to specify the amount of substance involved in the enthalpy change. If we attach the ΔH_{rxn} value to the chemical equation, we infer that the enthalpy change is associated with the number of moles indicated by the coefficients in the balanced equation. In this case, the equation is interpreted like this: When 1 mole of CH_4 reacts with 2 moles of O_2 to produce 2 moles of H_2O and 1 mole of CO_2, 890.4 kJ of heat is released into the surroundings. Chemists typically report the enthalpy changes for processes at a pressure of 1 atm, which they have agreed to call the **standard state.** For enthalpy changes at a pressure of 1 atm, a superscipt "o" is added to the ΔH_{rxn}. For example, the melting of ice at 1 atm is given by

$$H_2O(s) \longrightarrow H_2O(l) \qquad \Delta H° = 6.01 \text{ kJ}$$

(Here we omit the subscript rxn because it is a physical process.) In words this would read: When 1 mole of ice is converted to 1 mole of water at 0°C and 1 atm, 6.01 kJ of heat is absorbed from the surroundings.

Calculating Enthalpies

The enthalpy changes for many chemical processes have been measured experimentally, and these values can be used to calculate the enthalpy changes for many other processes. In Chapter 4, we said that chemical equations can be treated like algebraic equations. This means that you can add chemical equations together such that the same species that appear on both sides of the equation cancel each other and the same species on the same side of the arrow can be added together. For example, consider the following three reactions:

$$(1) \quad C(s) + \tfrac{1}{2}O_2(g) \longrightarrow CO(g)$$

$$(2) \quad CO(g) + \tfrac{1}{2}O_2(g) \longrightarrow CO_2(g)$$

$$(3) \quad C(s) + O_2(g) \longrightarrow CO_2(g)$$

If you imagine adding Equation (1) and (2) as though they were algebraic equations, the CO's would cancel each other, and the sum of the two equations would be Equation (3). **The importance of this additivity in thermochemistry is this: If two chemical equations can be added together to give a third, the enthalpy change of the third reaction is equal to the sum of the enthalpy changes of the first two reactions.** Because this rule has been proven to be true all of the time, it has achieved "Law" status. It is called the *law of heat summation* or, more commonly, **Hess's law,** after the 19th-century chemist Germain Hess. The law can be stated in a theoretical way (as Hess did) that does not involve adding equations (although it boils down to the same thing): When reactants are converted to products, the change in enthalpy is the same whether the reaction takes place in one step or in a series of steps.

Hess states his law of heat summation.

In our example, Equation (3) represents a reaction that can be broken down into two steps [(1) and (2)]. A useful analogy of Hess's law is as follows. Suppose you go from the first floor to the sixth floor of a building by elevator. The gain in your gravitational potential energy (which corresponds to the enthalpy change for the overall process) is the same whether you go directly there or stop at each floor on your way up (breaking the reaction into a series of steps). The algebraic additivity of thermochemical equations enables us to imagine any reaction as the sum of other reactions, which can be thought of as a series of reaction steps. Thus, we can calculate the enthalpy changes for reactions that may be difficult to study experimentally.

The Enthalpy of Elements Is Defined as Zero

As we said earlier, there is no way to measure the *absolute* value of the enthalpy of a substance. But understanding the way atoms, molecules, and ions behave requires having some idea of the instrinsic enthalpy (that is, thermal and chemical energy) of different substances. A way around the problem of not being able to determine absolute enthalpy values is to establish a reference point for enthalpies and assign enthalpy values relative to the reference. This problem is similar to the one geographers face in expressing the elevations of specific mountains or valleys. Rather than trying to devise some type of absolute elevation scale, by common agreement all geographic heights and depths are expressed relative to sea level, an arbitrary reference with a defined elevation of "zero" meter or feet. Similarly, we can set an arbitrary reference scale for the enthalpies of chemical substances. The "sea level" in our case (that is, zero enthalpy) is the enthalpy of all elements in their most stable form at 1 atm (the standard state); we arbitrarily define their enthalpy as zero.

The assignment of zero enthalpy is given to a specific form of an element because some elements (mostly the nonmetals) exist in different elemental forms. For example, oxygen can exist as O atoms, O_2 molecules, and O_3 molecules (ozone), but the most stable of these is O_2, so we say its enthalpy is zero. According to this reference point, the enthalpies of O and O_3 are greater than zero because heat energy is absorbed when they are made from diatomic oxygen. That is, it requires energy to make something less stable. Different forms of the same element are called **allotropes.** With the exception of the noble gases, the most stable allotrope of the nonmetals that are gases at room temperature is the diatomic molecule (e.g., H_2, N_2, F_2, Cl_2). The common allotropes of carbon are graphite and diamond. Graphite is the more stable allotrope. (If you wait long enough, you'd notice that a diamond ring changes into graphite. But this conversion may take millions of years to complete.) The most stable allotropes of some elements at 1 atm and 25°C are listed in Table 5.2.

Standard Enthalpy of Formation

With a reference scale established, we can experimentally determine the enthalpy change when compounds are formed from the stable form of their elements at standard state. For the formation of exactly 1 mole of a substance, we define a quantity called the **standard enthalpy of formation (ΔH_f°),** which is the heat change that results when 1 mole of a compound is formed from its elements at a pressure of 1 atm. The superscript "o" represents the standard-state condition of 1 atm, and the subscript f stands for formation. So ΔH_f° is the enthalpy change when a compound is formed from elements that we have defined as having zero enthalpy.

Let's say we are interested in knowing the ΔH_f° value of carbon dioxide. Carbon dioxide can be formed by burning elemental carbon in the form of graphite in an excess of air or oxygen:

$$C(\text{graphite}) + O_2(g) \longrightarrow CO_2(g)$$

Using an appropriate calorimeter, we find that the heat evolved is 393.5 kJ for every mole of CO_2 formed. Thus, ΔH_f° for carbon dioxide is -393.5 kJ per mole of CO_2 formed.

The ΔH_f° values for countless compounds have been determined directly by calorimetry, but there are many compounds for which we cannot determine ΔH_f° directly. For example, if we tried to measure the ΔH_f° value of carbon monoxide from graphite and molecular oxygen using a calorimeter, we would end up producing some carbon dioxide as well, even if we limited the amount of oxygen. However, we can

Table 5.2 Most Stable Allotropes of Some Common Elements

Element	Most Stable Allotrope
H	$H_2(g)$
C	C(graphite)
N	$N_2(g)$
O	$O_2(g)$
F	$F_2(g)$
P	$P_4(s)$
S	$S_8(s)$
Cl	$Cl_2(g)$
Br	$Br_2(l)$
I	$I_2(s)$

calculate the ΔH_f° of CO if we use Hess's law. We can consider the formation of CO as the first step in the overall formation of CO_2 and the reaction of CO and O_2 to form CO_2 as the second step. The ΔH_{rxn}° for the latter reaction can be measured by calorimetry, which gives a value of -283.0 kJ per mole of CO_2 formed. The thermochemical equations can be summed as follows:

$$(1)\ \ C(graphite) + \tfrac{1}{2}O_2(g) \longrightarrow CO(g) \qquad\qquad \Delta H_{rxn}^\circ = \ ?$$

$$(2)\ \ CO(g) + \tfrac{1}{2}O_2(g) \longrightarrow CO_2(g) \qquad\qquad \Delta H_{rxn}^\circ = -283.0\ kJ$$

$$(3)\ \ C(graphite) + O_2(g) \longrightarrow CO_2(g) \qquad\qquad \Delta H_{rxn}^\circ = -393.5\ kJ$$

As you have seen before, Equation (3) is the sum of (1) and (2). Therefore, according to Hess's law, we can calculate the ΔH_{rxn}° for reaction (1) by writing

$$\Delta H_{rxn}^\circ\ (3) = \Delta H_{rxn}^\circ\ (1) + \Delta H_{rxn}^\circ\ (2)$$

or
$$\Delta H_{rxn}^\circ\ (1) = \Delta H_{rxn}^\circ\ (3) - \Delta H_{rxn}^\circ\ (2)$$

$$= -393.5\ kJ - (-283.0\ kJ)$$

$$= -110.5\ kJ$$

Because Equation (1) represents the standard enthalpy of formation for CO,

$$\Delta H_{rxn}^\circ = \Delta H_f^\circ\ of\ CO = -110.5\ kJ/mol$$

5.1

EXAMPLE

Calculate the standard enthalpy of formation of methane (CH_4) from its elements:

$$C(graphite) + 2H_2(g) \longrightarrow CH_4(g)$$

Given that

$$(1)\ \ C(graphite) + O_2(g) \longrightarrow CO_2(g) \qquad\qquad \Delta H_{rxn}^\circ = -393.5\ kJ$$

$$(2)\ \ 2H_2(g) + O_2(g) \longrightarrow 2H_2O(l) \qquad\qquad \Delta H_{rxn}^\circ = -571.6\ kJ$$

$$(3)\ \ CH_4(g) + 2O_2(g) \longrightarrow CO_2(g) + 2H_2O(l) \qquad \Delta H_{rxn}^\circ = -890.4\ kJ$$

Note that Equation (1) is the standard enthalpy of formation of CO_2 and Equation (2) corresponds to twice the standard enthalpy of formation of H_2O because 2 moles of water are formed (ΔH_f° of water is -285.8 kJ/mol).

Answer   *We cannot prepare CH_4 by heating graphite in hydrogen because many different products will be formed. So the only way to do this problem is to apply Hess's law. The general rule you should keep in mind is that Equations (1), (2), and (3) must be rearranged such that the reactants (graphite and H_2) appear on the left and the product (CH_4) appears on the right of the arrow. By proper manipulation, the other substances (CO_2, O_2, and H_2O) will cancel out. We note that graphite and H_2 are already on the left but CH_4 is also on the left. So, we reverse Equation (3) to get*

$$(4)\ \ CO_2(g) + 2H_2O(l) \longrightarrow CH_4(g) + 2O_2(g) \qquad \Delta H_{rxn}^\circ = +890.4\ kJ$$

—Continued next page

Continued—

(Remember that when we reverse a reaction, the sign of ΔH°_{rxn} must change. This means that an exothermic reaction becomes endothermic, and vice versa.) Now we add Equations (1), (2), and (4):

(1) $C(graphite) + O_2(g) \longrightarrow CO_2(g)$ $\Delta H^{\circ}_{rxn} = -393.5 \text{ kJ}$

(2) $2H_2(g) + O_2(g) \longrightarrow 2H_2O(l)$ $\Delta H^{\circ}_{rxn} = -571.6 \text{ kJ}$

(4) $CO_2(g) + 2H_2O(l) \longrightarrow CH_4(g) + 2O_2(g)$ $\Delta H^{\circ}_{rxn} = +890.4 \text{ kJ}$

$\overline{C(graphite) + 2H_2(g) \longrightarrow CH_4(g)}$ $\Delta H^{\circ}_{rxn} = -74.7 \text{ kJ}$

All the unwanted substances cancel on both sides of the equations. Thus, ΔH°_f of methane is -74.7 kJ/mol.

Table 5.3 Standard Enthalpies of Formation for Some Compounds

Compound	ΔH°_f (kJ/mol)
$CO(g)$	-110.5
$CO_2(g)$	-393.5
$H_2O(l)$	-285.8
$NH_3(g)$	-46.3
$NO_2(g)$	33.9
$CH_4(g)$	-74.7
$C_6H_6(l)$	49.0
$C_6H_{12}O_6(s)$	-1261

Calculating ΔH°_{rxn} from ΔH°_f

Over the years, chemists have compiled a large body of ΔH°_f data of various compounds. Standard enthalpies of some common compounds are listed in Table 5.3. These data provide us with a way to calculate the ΔH°_{rxn} of a reaction without doing an experiment. To see how this works, let's apply Hess's law to the production of glucose ($C_6H_{12}O_6$) and oxygen from carbon dioxide and water that occurs in plants:

$$6CO_2(g) + 6H_2O(l) \longrightarrow C_6H_{12}O_6(s) + 6O_2(g)$$

This chemical reaction is the basis for photosynthesis, in which radiant energy from the sun drives this highly endothermic process. (Solar energy produces about 7.0×10^{14} kg of glucose each year on Earth.) As you can imagine, it would not be easy to study this reaction in a calorimeter. According to Hess's law, we can calculate ΔH°_{rxn} for the photosynthetic reaction if we break it down into several reactions of known ΔH°_{rxn} that add up to the formation of glucose and oxygen from carbon dioxide and water. Using the ΔH°_f values for the products and reactants is a convenient way to do this. Because these values reflect the enthalpy of a compound relative to the arbitrarily defined zero enthalpy state for elements in their most stable form at standard state, it's like comparing the heights of mountains based on their altitudes relative to sea level.

We can consider any reaction as a series of indirect chemical steps: first, the breakdown of the reactants to the elements in their most stable form at standard state, then the formation of the products from the elements in their most stable form at standard state. In this case, the first two steps involve the ΔH°_f for the reactants written in reverse because the compounds are broken down to the elements.

(1) $6CO_2(g) \longrightarrow 6C(graphite) + 6O_2(g)$

(2) $6H_2O(l) \longrightarrow 6H_2(g) + 3O_2(g)$

The next two steps correspond to the ΔH°_f of the products from their elements.

(3) $6C(graphite) + 6H_2(g) + 3O_2(g) \longrightarrow C_6H_{12}O_6(s)$

(4) $6O_2(g) \longrightarrow 6O_2(g)$

By definition, the ΔH°_f of $O_2(g)$ is zero, so we can ignore Equation (4). Note that addition of the other three equations places everything on the side of the arrow that we want in the final equation. However, in order for the equations to add up correctly, we need 6 moles of CO_2 and 6 moles of H_2O, so we will have to multiply the ΔH_f° values by 6. We must also reverse the signs of these enthalpy changes because we have written the standard enthalpy of formation equations for CO_2 and H_2O in reverse. The thermochemical equations would be written as follows (see Table 5.3):

(1) $6[CO_2(g) \longrightarrow C(graphite) + O_2(g)]$ $-\Delta H_f^\circ = 6 \times 393.5 \text{ kJ} = 2361 \text{ kJ}$

(2) $6[H_2O(l) \longrightarrow H_2(g) + \frac{1}{2}O_2(g)]$ $-\Delta H_f^\circ = 6 \times 285.8 \text{ kJ} = 1715 \text{ kJ}$

(3) $6C(graphite) + 6H_2(g) + 3O_2(g) \longrightarrow C_6H_{12}O_6(s)$ $\Delta H_f^\circ = -1261 \text{ kJ}$

$6CO_2(g) + 6H_2O(l) \longrightarrow C_6H_{12}O_6(s) + 6O_2(g)$ $\Delta H_{rxn}^\circ = 2815 \text{ kJ}$

As you may have realized, we don't really have to go through all these steps because we have simply subtracted the enthalpies of formation of the reactants from the enthalpies of formation of the products. In fact, for any chemical reaction, ΔH_{rxn}° can be calculated using the formula

$$\Delta H_{rxn}^\circ = n\Delta H_f^\circ(\text{products}) - n\Delta H_f^\circ(\text{reactants})$$

where n represents the number of moles of each substance (that is, the stoichiometric coefficients) in the balanced equation. This formula also illustrates that you can easily predict whether a reaction will be endothermic or exothermic by comparing the magnitudes and signs of the enthalpies of formations. If ΔH_f° of the products is less than that of the reactants, the reaction will be exothermic; if ΔH_f° of the products is greater than that of the reactants, the reaction will be endothermic.

Understanding Enthalpy Changes

While Hess's law and ΔH_f° values are very useful for calculating enthalpy changes in chemical reactions, they don't shed any light on the nature of enthalpy. To understand the meaning of enthalpy changes for chemical processes, it's important to think about the changes in the atoms, molecules, and ions as a result of the reaction. Let's consider how we might do this for the photosynthetic reaction described earlier. First of all, a value of 2815 kJ for ΔH_{rxn}° tells us that the reactants are more stable than the products (that is, the enthalpy, or energy, of the reactants is less than that of the products). We infer that the combined chemical forces holding the atoms and ions together in the reactants are greater than that of the products. In fact, comparison of the enthalpies of formation of the reactants and products shows that the combined forces holding the former together are much stronger. What are these forces?

The forces that hold atoms, molecules, and ions together were described earlier in the chapter when we discussed chemical energy. In the photosynthetic reaction, most of the enthalpy change is due to changes in chemical energy. If we break down the reaction into the steps that must occur in the overall process, the nature of all of the forces that give rise to the change in chemical energy is apparent. To imagine these steps, you need to recognize that all of the species are molecules (all nonmetals, remember?) held together by covalent bonds, which are either broken or formed during the reaction. You also need to consider that the water and glucose molecules are in the liquid and solid states, respectively, and are therefore held together by intermolecular forces. Now, if we think about the energy involved, the first thing that has to happen is this: Intermolecular forces among water molecules are broken as are the O—H covalent bonds in water and the C=O covalent bonds in carbon dioxide. The energy used to break these bonds is provided by radiant energy from the sun. Conversely, energy is released when all the covalent bonds form between C, H, and O in

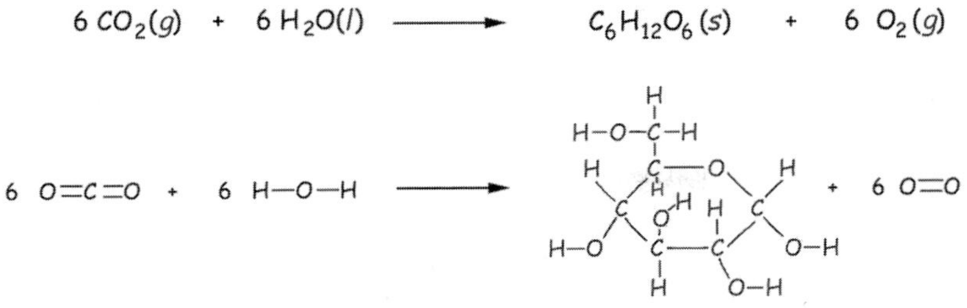

$$6\ CO_2(g)\ +\ 6\ H_2O(l)\ \longrightarrow\ C_6H_{12}O_6\ (s)\ +\ 6\ O_2\ (g)$$

$$6\ O{=}C{=}O\ +\ 6\ H{-}O{-}H\ \longrightarrow$$

(glucose structural formula) $+\ 6\ O{=}O$

BONDS BROKEN		BONDS FORMED	
CARBON DIOXIDE:	12 C=O BONDS	GLUCOSE:	5 C—C BONDS
			7 C—O BONDS
WATER:	12 O—H BONDS		7 C—H BONDS
			5 O—H BONDS
		OXYGEN:	6 O=O BONDS

Figure 5.2 The enthalpy change for the photosynthetic formation of glucose can be estimated using bond energies. Determine the total energy needed to break all the bonds in the reactants and the total energy released by forming bonds in products using Table 5.3. Subtraction of the energy of bond breaking from the energy released is the enthalpy change.

Table 5.4 Bond Energies for Some Common Single, Double, and Triple Bonds

Bond Type	Bond Energy (kJ/mol)
C—H	414
C—O	351
C=O	799
C—C	347
C=C	620
C≡C	812
C—N	276
C=N	615
C≡N	891
N—O	176
N=O	458
O—H	460
O=O	499
H—F	568
H—Cl	432
H—Br	366
H—I	298

glucose and between O atoms in molecular oxygen. Additional energy is released when glucose molecules stick together in the solid state.

The energy required to break covalent bonds (called bond energy) has been determined for many compounds (Table 5.4). Let's return to the photosynthetic reaction. Figure 5.2 lists all the covalent bonds that are broken or formed in the reaction and the combined energy associated with breaking or making them. (Glucose is a bigger and more complicated molecule than those we have considered so far, but note that every C forms four bonds, every O forms two bonds, and every H forms one bond as you already know.) Adding these bond energy values gives a ΔH°_{rxn} of 2724 kJ for the reaction. Note that this is 91 kJ lower than the value we obtained using ΔH°_f values of the reactants and products. One reason for the different values is that we did not consider the energy associated with breaking and forming intermolecular forces in water and glucose. However, a more likely explanation is that the bond energy values are approximate values obtained by averaging the values for a variety of different compounds. *(Check out the website for more problems.)*

Summing Up Energy Changes

Now that we've covered all of the theoretical and quantitative aspects of thermochemistry you should know, let's summarize the major points you must remember. (1) Chemical reactions are accompanied by changes in energy, which can be radiant, thermal, or chemical. (2) Energy is conserved, so the energy released by a system is absorbed by the surroundings, and vice versa, according to the law of conservation of energy. (3) Based on this law, energy changes can be measured by calorimetry. (4) We define the energy change determined by calorimetry (at constant P) as enthalpy. (5) The enthalpy change for a process is the same if it happens in one step or several steps: Hess's law. (6) Based on Hess's law, we can calculate enthalpy changes for processes that are difficult or impossible to measure directly; that is, breaking down the overall process into a number of steps. (7) We can't measure absolute enthalpies of substances, but we can determine relative enthalpies of substances by defining a reference point of zero enthalpy for all elements in their stable allotropic form at standard state (that is, 1 atm and 25°C).

Test your understanding of the material in this chapter

IMPORTANT TERMS

Explain the following terms in your own words:

allotropes, p. 73

calorimeter, p. 67

calorimetry, p. 67

covalent bond, p. 63

endothermic
  process, p. 66

energy, p. 61

enthalpy, p. 70

exothermic process, p. 66

heat, p. 63

Hess's law, p. 72

intermolecular
  forces, p. 64

ionic bond, p. 64

law of conservation of
  energy, p. 65

specific heat, p. 68

standard enthalpy of
  formation, p. 73

standard state, p. 71

thermochemistry, p. 65

CONCEPTS

Explain in your own words:

- Different types of energy and their interconversions.
- The measurement of heat change using a constant-pressure calorimeter.
- Determining the standard enthalpy of formation of a compound.
- Determining the enthalpy change of a reaction using the standard enthalpies of formation of products and reactants.

Atomic Structure and the Periodic Table

Peeking into the atom

I n Chapter 2 we introduced you to the composition of atoms. You know they contain protons, neutrons, and electrons. You know from Rutherford's experiments that the density of the nucleus is huge and that most of the space in an atom is occupied by the lightweight electrons. You also know that it's the electrons that are involved in chemical reactions—they're either transferred or shared. In order to understand why and how electrons are transferred or shared by atoms, you will need to learn something about their behavior. What exactly are those electrons doing when they're moving around the nucleus? First of all, we should tell you that no one fully understands electrons, but we know a lot about where they spend their time and how they behave. It's like gravity; nobody fully understands it, but we know how it works.

Knowing the location and behavior of electrons within an atom is extremely useful because it forms the basis for predicting chemical properties of substances. Moreover, because atoms are grouped according to chemical properties in the periodic table, the location and behavior of electrons are inherent in the layout of the periodic table—hence, the title of this chapter. Our closer look at atoms in this chapter is really a closer look at the electrons in atoms and how their location and behavior relate to the periodic table. To set the stage for our discussion of the properties of electrons, we'd like to give you a little history about the properties of radiant energy (e.g., light) because electrons and light behave similarly, although light is a bit easier to understand.

Light

In Chapter 5 we told you that **hot objects emit light.** When you turn on a lightbulb, it emits radiant energy because the tungsten filament is heated to about 3000°C by electrical energy. Experiments have shown that radiant energy is transmitted through space in periodic waves, not unlike sound or the waves that are transmitted through water.

Light Is a Wave

To understand that light travels as a wave, you have to understand the basic properties of waves. Imagine two people holding the ends of a stretched piece of string. If one person moves his end up and down quickly, a pulse is sent along the string—this is a wave. If the end is repeatedly moved up and down, a periodic wave is generated. Now if both people move their ends by the same amount at the same time such that the waves overlap when they meet, the size of the wave, or the displacement of the string, at this point is doubled. These waves are said to be in phase with one another. On the other hand, if the waves are generated from each end such that they are completely out of phase (that is, the hump of one overlaps with the trough of the other), there is no string displacement where they meet.

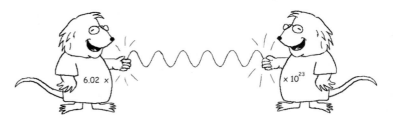

The ability of waves to interfere with one another *constructively* when they are in phase and *destructively* when they are out of phase provided the basis for the proof that light has wave properties. In 1801 the English scientist and physician **Thomas Young** performed the classic double-slit experiment in which he shined light against a surface through two adjacent slits and observed a pattern of alternating bright and dark lines. (Try this out at home with a flashlight.) If the light is shined through only one slit, the entire surface is lighted. Young explained his results by saying that the dark lines were produced when the light from the two sources interfered with each other destructively and the bright lines were due to constructive interference—exactly what you'd expect if light travels in waves.

Young saw light . . .
and proved it was a wave.

Waves can be represented very simply as shown in Figure 6.1 by oscillating lines and can be described by two features—wavelength (λ) and frequency (ν). One wave-

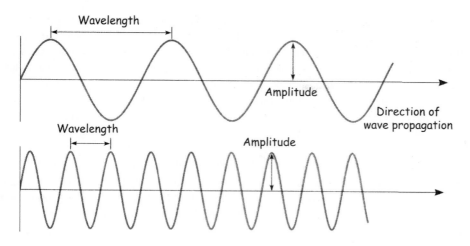

Figure 6.1 Two waves having the same amplitude but different wavelengths and frequencies

length can be thought of as the horizontal displacement in one complete oscillation. Note that the two waves in Figure 6.1 differ in their length, with the one on the top being three times as long as the one on the bottom. Radiant energy is transmitted in an infinitely large number of different wavelengths ranging from 10^{-14} m for gamma rays to 1000 m for radio waves. If you hold a prism in front of a lightbulb, you will see all the colors of the rainbow (that is, the entire spectrum of visible light). The prism separates the light into its component colors, which correspond to the many distinct wavelengths of light. For example, blue light has a wavelength of about 4×10^{-7} m, or 400 nm (1 nm = 1×10^{-9} m; see back endpaper for prefixes), and red light has a wavelength nearly twice as long (about 700 nm). The frequency of a wave is the number of oscillation cycles per second and has the unit of s^{-1}. If the waves in Figure 6.1 were propagated at the same speed, the frequency of the wave on the bottom would be three times that of the wave on the top. **Waves with short wavelengths have a high frequency.** The speed at which a wave travels is the product of the wavelength times the frequency (m $\times$ s^{-1} = m/s). This should make sense if you think about the oscillating string. Let's say the string oscillates at a frequency of one oscillation per second (that is, the string goes up and down and returns to its original position in one second). If the length of the wave were 10 cm, the wave will travel 10 cm per second, which is its speed. An important point about radiant energy is that all radiation waves travel at the same speed—the speed of light (3.0×10^8 m/s). **For all radiant energy, the product of the wavelength (in meters) times the frequency (in s^{-1}) is 3.0×10^8 m/s in a vacuum.** For our purposes, we can assume the speed of light is a constant and assign the symbol c for it, so its relationship to wavelength and frequency is given by the equation: $c = \lambda \nu$.

But what exactly is traveling in these waves of radiant energy? Light is not a tangible substance like matter, but it is measurable. Imagine our lightbulb is on a dimmer switch. If you adjust the switch so that the light is about half as bright—midway between completely on and off—you would expect the energy emitted would be half as much. But, if you were to hold up your prism to the dimmed light, all the same colors (that is, wavelengths) would be there. We haven't changed the speed, wavelength, or frequency, but the radiant energy given off is obviously less. To understand how this might work, it's useful to think about reducing the amount of matter as an analogy. Remember in Chapter 2 we said you can go only so far in dividing a piece of gold wire. When you reach the size of a single atom, you can't divide it anymore and still have gold. Imagine there is a similar point for radiant energy, the smallest unit of radiant energy, where you can't reduce the energy anymore (by dimming the light) and still have radiant energy.

This turns out to be the case, and we owe our understanding of this property of radiant energy to two German physicists who pioneered what we call **quantum theory**—*Max Planck* and *Albert Einstein.* Because quantum theory can't be explained in a way that makes sense in our view of the "macroscopic real" world, it is best understood by following its development from the beginning. It all started when these two brilliant scientists proposed a radical idea to explain experimental results with light that could not be explained by classical physics. The experiments dealt with two light-related phenomena called *black-body radiation* and the *photoelectric effect.*

Planck Explains Black-Body Radiation: Energy Is Quantized

It has been known for a long time that hot objects emit light. Hundreds of years ago, potters used color to determine the temperature of their hot kilns; red is 700°C and yellow is 1000°C (no matter what the kiln is made of). Experiments at the end of the 19th century dealing with so-called black-body radiation involved measuring the frequency of light emitted from a black object (or body) when it's heated. Things are black because they absorb all the radiation that falls on them. Things that are colored don't absorb all wavelengths of radiation, but reflect back the colors we see; white objects reflect all of the visible light. So using a black object (or body) ensures that a maximum number of visible wavelengths are absorbed. The apparatus used was a lot like a black oven with a slit for light to come out. The oven was heated, and the frequencies and intensity of light coming from the slit were measured. When the intensity of the light was plotted against wavelength, the data looked like bell-shaped curves that were bigger and shifted toward shorter wavelengths at higher temperatures.

Early attempts to derive equations for the curve based on classical physics approaches failed to account for all the data. For example, if you considered that the waves simply got more intense as their wavelength decreased, the curve would keep going up indefinitely (and short-wavelength light could be very dangerous). In 1900 Max Planck tried his hand at explaining the black-body radiation results that confounded the scientific community and colleagues. He proposed that the energy radiated from a black body is not continuous, but exists as discrete energy packets whose energy is proportional to its frequency ($E \propto \nu$). He called the packets **quanta** and determined the proportionality constant (h, now called Planck's constant) to be a very small number, 6.626×10^{-34} J s, ($E = h\nu$). Thus, radiant energy is said to be *quantized;* that is, it is limited to amounts that are whole number multiples of

RADIANT ENERGY EXISTS AS DISCRETE PACKETS, AND THE ENERGY OF ONE OF THESE PACKETS IS PROPORTIONAL TO ITS FREQUENCY: E =hv.

Planck proposes that radiant energy is quantized.

the basic unit $h\nu$ (1 $h\nu$, 2 $h\nu$, 3 $h\nu$, . . .). If we invoke the matter analogy, $h\nu$ is to energy what the atom is to matter, and just as there are different types of elements, radiant energy comes in different colors and forms. The energy of a quantum of blue light would be higher than that of red light because blue light has a higher frequency. Although most things we are familiar with are quantized—from atoms to people (you can't divide either of them), it had never before been considered for something with wave properties. While Planck's equation fit the experimental data perfectly, neither he nor anybody else really believed it for a long time. The idea of quantization of energy, that is, energy changes only in discrete amounts, will take you a while to get used to.

Einstein Explains the Photoelectric Effect: Light Is a Particle

A few years later, Einstein elaborated on Planck's quantum theory by proposing that radiant energy is emitted in discrete units, or light particles (called photons). The basis for his proposal was the so-called photoelectric effect, the experimental observation that electrons can be ejected from a metal by light if the frequency of the light is high enough. For example, you can shine intense red light on potassium metal in a vacuum for hours without ejecting a single electron, but as soon as you shine yellow light (shorter wavelength, higher frequency and higher energy) on the metal, electrons are easily knocked out of the atoms. If you use blue light, the electrons are also ejected, but with a lot more kinetic energy. This behavior perplexed physicists because the classical view held that even low-energy red light should eject electrons if the intensity of the light were high enough.

Einstein explained the phenomenon as follows: The removal of an electron from the metal requires a finite amount of energy. When a photon having enough energy ($h\nu$) collides with an atom in the metal, it transfers all of its energy to the most loosely held electron. If the energy of the photon is higher than the amount needed to knock the electron out of the atom, the extra energy is absorbed by the electron in the form of kinetic energy. Every metal has a threshold frequency of radiation below which an electron cannot be removed no matter how intense the light. The notion that light possesses both wave and particle properties was a very radical idea at the time, and few scientists bought into it. The explanation was consistent with Planck's notion of energy quanta and eventually proved to be correct by a classical physicist, ***Robert Millikan,*** who conducted painstaking experiments trying to prove Einstein wrong.

Einstein explains the
photoelectric effect.

J.J. Thomson and Lord Kelvin
discuss atomic structure.

Electrons

When Einstein proposed his idea of light particles transferring their energy to electrons in atoms, a model of atomic structure had been proposed by ***J.J. Thomson*** (who discovered the electron and showed it to be a particle with mass and charge) and ***Lord Kelvin.*** They suggested an atom was a positive sphere with negative electrons embedded in it, like plums in pudding. A few years later, Rutherford's experiments with alpha particles (recall from Chapter 2) led him to propose his model of a mostly empty atom where electrons moved around the very small, positive nucleus, like planets around the sun. But there were problems with this view of electrons. According to classical physics, the electron should eventually collapse into the nucleus. A major breakthrough in our understanding of electrons came when the Danish physicist ***Neils Bohr*** developed a model to explain the emission spectrum of gases, another seemingly inexplicable experimental result. In so doing, Bohr extended quantum theory to the realm of the atom.

Bohr Explains Emission Spectra: The Energies of Electrons Are Quantized

Since the middle of the 18th century, scientists have studied the emission of light from heated gases. Unlike the emission from a lightbulb, which can be separated into a continuous spectrum of visible wavelengths with a prism, the emission of hot gases through a prism gives a small number of distinct lines at specific wavelengths. It's as if large chunks of the visible spectrum were blacked out, as shown in Figure 6.2. In 1885 a Swiss mathematician, ***Johann Balmer,*** discovered a simple formula that accounted for the frequencies of the distinct colored lines of the hydrogen spectrum that involved small whole numbers:

$$\nu = R\left(\frac{1}{n_f^2} - \frac{1}{n_i^2}\right)$$

For the four visible lines in the hydrogen spectrum, the value of n_f (for n_{final}) is 2, and the values of n_i (for $n_{initial}$) are 3, 4, 5, and 6. He determined the value of R, called the Rydberg constant, to be 3.29163×10^{15} s^{-1}. This remarkably simple equation gives the frequencies of the visible spectral lines to an amazing degree of accuracy and predicts many other frequencies outside the visible region.

Bohr learned about Balmer's equation at a time when he was already thinking about the arrangement of electrons in an atom. Inspired by Rutherford's atomic model

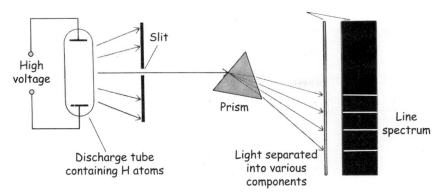

Figure 6.2 Emission spectrum of hydrogen atoms

and the quantum ideas of Planck and Einstein, he developed a model of the hydrogen atom that would account for the hydrogen emission spectrum. He reasoned that the spectral lines represent photons with specific energies that are emitted when an electron loses the energy it gained when the gas was heated. Accordingly, the electron must be able to exist for some finite period of time in different energy states within the atom. A high-energy-state electron will eventually want to lose energy to return to its most stable lower-energy state, called the **ground state** or **ground level.** An electron having more energy than its most stable state is said to be in an **excited state** or **excited level.**

THE ENERGY LEVELS OF ELECTRONS ARE QUANTIZED. HYDROGEN'S ELECTRON CAN BE MOVED INTO HIGHER ENERGY LEVELS IF IT ABSORBS ENERGY FROM A PHOTON WITH JUST THE RIGHT AMOUNT OF ENERGY.

Bohr proposes a model for the hydrogen atom based on its emission spectrum.

Bohr proposed that the emission from hydrogen atoms was restricted to certain frequencies because the lone electron can possess only certain energies, each of which corresponds to a particular orbit around the nucleus. In other words, the energy of the single electron orbiting the nucleus in the H atom is quantized. A useful analogy for Bohr's hydrogen atom is a multistory building where the floors (energy levels) can be reached only by an electron using the elevator. The ground floor corresponds to the ground state, and the upper floors are excited states. Energy must be supplied to raise the elevator to the higher floors, and energy is released when the elevator goes down. The elevator can remain at a particular floor momentarily (but not between floors). Now, if the energy were released in the form of photons, a drop from the 10th floor to the ground floor would produce a photon with higher frequency (and lower wavelength) than the photon emitted in going from the second floor to the ground floor.

Figure 6.3 shows a diagram of the Bohr model. In the model, an electron emits a photon when it drops from a higher-energy orbit (an excited state) to a lower-energy

Figure 6.3   According to the Bohr model of the hydrogen atom, the electron can be promoted to a higher energy level by absorbing a discrete amount of energy. The electron will then drop to a lower energy level, emitting a photon that has a specific wavelength corresponding to the energy change.

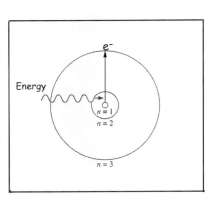
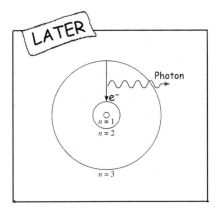

orbit (a less excited state or the ground state). This model accounts for the specific energies represented by the lines in the hydrogen emission spectrum. In a broader sense, the model also suggests an explanation for the light emitted from hot objects: The electrons are excited by the heat and emit photons when they drop to lower-energy states. Because many more energy states are available in solids with lots of electrons, their spectrum appears to contain all frequencies of visible light.

However, one of the major shortcomings of the Bohr model was that it could not explain the emission spectra of atoms with more than one electron (not even the helium atom). The fundamental problem with the model is that electrons do not circle the nucleus in predictable paths or orbits. Describing the behavior of electrons would require thinking about electrons in a different way altogether.

Electrons Are Waves

In 1924 a graduate student in Paris, Prince **Louis de Broglie,** astounded his Ph.D. review committee when he presented a thesis (about three pages in length) on the wave properties of particles. Using photons as a model, he started with Einstein's famous mass-energy relationship equation: $E = mc^2$, where c is the speed of light. He then substituted the momentum, p (which is mass times velocity):

$$E = pc = p(\lambda \nu) \qquad \text{(remember } c = \lambda \nu)$$

Because $E = h\nu$,

$$h\nu = p(\lambda \nu)$$

then

$$\lambda = h/p \text{ or } \lambda = h/mu$$

where λ is the wavelength, *m* the mass of the system, and *u* the velocity. de Broglie proposed that this relationship applies to all matter in motion, from electrons to tennis balls. That is, a moving object oscillates with a wavelength that depends on both its velocity and its mass—the larger the mass or the velocity, the smaller the wavelength. This was such a radical idea that the scientists who reviewed de Broglie's thesis sent a copy to Einstein for his opinion. Impressed with de Broglie's work, Einstein told them to give de Broglie his degree. **So both light and electrons behave as waves and particles (sometimes called wave-particle duality), and their energies are quantized.**

ELECTRONS BEHAVE LIKE WAVES. YOU CAN READ ALL ABOUT IT IN MY PH.D. THESIS IF YOU HAVE A MINUTE.

de Broglie's Ph.D. thesis may be the shortest in history.

de Broglie's theory provides the needed explanation for the quantized energy states of electrons in atoms. Consider a circular orbit like the one shown in Figure 6.4. If the circumference of the orbit is equal to a whole number multiple of a certain wavelength, an electron having that wavelength (and the corresponding energy) is allowed. Thus, an electron with a wavelength as shown on top is allowed. However, an electron could not have the energy corresponding to the wavelength in the figure on the bottom because in going around the orbit, the waves would be out of phase and therefore couldn't exist. Ironically, the wavelike nature of electrons was later proven by *G.P. Thomson,* 30 years after his father (J.J. Thomson) showed that electrons are particles.

Electrons Inhabit Orbitals

Inspired by de Broglie's notion of the wave properties of electrons, the Austrian physicist *Erwin Schrödinger* derived an equation in 1926 that describes electrons as waves in three-dimensional space. The Schrödinger wave equation is the basis of **quantum mechanics,** which describes the behavior of electrons as effectively as Newtonian mechanics describes the orbits of planets around the sun. It is not an easy subject. Bohr reportedly said that if you weren't confused by quantum mechanics, then you didn't really understand it. Understanding quantum mechanics is well beyond the scope of an introductory chemistry course, but you can understand what it tells us about atomic structure. If the Schrödinger equation is solved for the hydrogen atom, which requires some advanced calculus, you get a set of so-called *wave functions* with energies that agree with the energies calculated by Bohr. In principle, this equation works for multielectron atoms and molecules as well.

The most important outcome from solving the Schrödinger equation for atoms is the description of **orbitals.** Atomic orbitals come in different sizes, shapes, and orientations. They are well-defined regions of three-dimensional space that can be inhabited

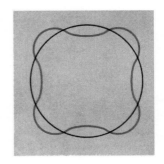

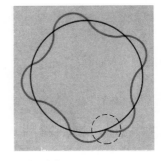

Figure 6.4 Allowed *(top)* and nonallowed *(bottom)* orbits

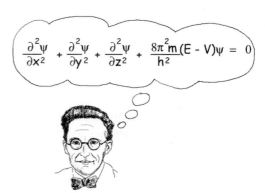

Schrödinger's equation describes
electron behavior in atomic orbitals.

by an electron, but it's not possible to know exactly where the electron will be in that space at a particular time, just like it is not possible to describe the "location" of a wave in space. We just know that it will most probably (better than a 90% chance) be somewhere within the boundaries of the orbital space. If you could take a movie of the electron and overlap all of the frames (that is, average its location over time), it would look like a cloud of dust distributed throughout the orbital. In fact, electron orbitals are often referred to as electron clouds or electron densities. Because orbitals come in different sizes, shapes, and orientations, the location of any electron inhabiting an orbital could be identified by assigning specific designations for these three parameters. Solving the Schrödinger equation gives three **quantum numbers,** which do just that.

Quantum Numbers Describe Orbitals

To understand the meaning of quantum numbers, let's consider another building analogy. In this case, the atom is a hotel with several rooms of different types on each floor—some regular rooms, some suites, and some deluxe suites. Depending on the atom, the building could be filled with any number of electrons, from one (like hydrogen) to one in every room. A particular electron can be found in the building if you have an address identifying the specific room. For example, to identify a third-floor room, you might give it the address 3SW, where 3 refers to the floor, S stands for suite, and W means it's on the west side of the building.

Similarly, an electron in an orbital can be identified by a distinct set of three quantum numbers, one to indicate the orbital's energy level (or size, as in the Bohr model), another to indicate its shape, and a third to indicate its spatial orientation. A fourth quantum number was proposed by the Swiss physicist *Wolfgang Pauli.* Pauli suggested that each orbital can hold two, and only two, electrons, each one spinning on it's own axis in clockwise and counterclockwise directions. Experimental evidence has since shown this to be correct. Pauli introduced the spin of an electron as a fourth quantum number, and his idea was extended to the assignment of quantum numbers as the **Pauli exclusion principle: No two electrons in an atom can have the same four quantum numbers.**

The three values that come from Schrödinger's wave equations are called the principal (n), angular momentum (l), and magnetic (m_l) quantum numbers. They designate the size, shape, and special orientation of the orbitals in an atom. The combination of these three numbers gives the address of an electron. For example, an address of 211 would correspond to $n = 2$, $l = 1$, and $m_l = 1$. The possible values for each quantum number and their meaning are given by some simple rules.

AN ORBITAL CAN HOLD A MAXIMUM OF TWO ELECTRONS, WHICH SPIN AROUND IN OPPOSITE DIRECTIONS.

Pauli's exclusion principle:
No two electrons in an atom can
have the same four quantum numbers.

n

The **principal quantum number** (n) describes the size and energy level of the orbital and can have integral values of 1, 2, 3, and so on, where higher numbers mean bigger size of the orbital and higher energy. It's analogous to the energy levels in Bohr's hydrogen atom. For example, orbitals for which $n = 2$ are larger and have higher energy than those for which $n = 1$ (which is the ground state). The principal quantum number is the first number in the electronic address, so any electron in the $n = 3$ level, for example, will have 3 as its first number in the address. As with the Bohr model, energy must be absorbed to excite an electron from a lower-energy orbital to a higher one. The different energy levels in atoms are often called *shells,* where each n value corresponds to a distinct shell.

l

The **angular momentum quantum number** (l) describes the shape of the orbital and can have integral values of 0, 1, 2, 3, and so on; however, the value of l depends on the value of n. For a particular value of n, l can have values only up to $n - 1$. Figure 6.5 shows the shapes of the so-called s, p, and d orbitals; these orbital types have l values of 0, 1, and 2, respectively. Orbitals come in more complex shapes (and too hard to draw) as the value of l gets larger, but we don't need to concern ourselves with the shape of anything above $l = 2$, which is the angular momentum quantum number for d orbitals. However, in theory, they could go on . . .

l	0	1	2	3	4	5
Name of orbital	s	p	d	f	g	h

At each energy level (n), there can be one or more types of orbitals. For example, at $n = 1$, $l = 0$, so only s orbital exists; at $n = 2$, $l = 0$ and 1, so s and p *orbitals* are present; at $n = 3$, $l = 0$, 1, or 2, so s, p, and d orbitals are present; and at $n = 4$, $l = 0$, 1, 2, or 3, so s, p, d, and f orbitals are all present. As the n number increases,

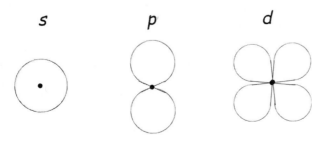

s   *p*   *d*

Figure 6.5 Shapes of orbitals. The nucleus is indicated by a black dot.

the number of types of orbitals increases as well. The value of l and orbital names (which are simply letters) can go on indefinitely, but the only ones that are relevant to us at this level are the *s, p, d,* and *f* orbitals.

As mentioned previously, a collection of orbitals with the same value of n is called a shell. One or more orbitals with the same n and l values (that is, the same orbital type) are referred to as *subshells.* For example, the shell with $n = 2$ is composed of two subshells, $l = 0$ and 1 (the allowed values for $n = 2$). These are the 2s and 2p subshells, where 2 denotes the value of n, and s and p correspond to $l = 0$ and 1.

m_l

The **magnetic quantum number** (m_l) describes the orientation of the orbital in space and has values that go from $-l$ to 0 to $+l$. It's called the *magnetic* quantum number because the effect of different orbital orientations was first observed in the presence of a magnetic field. Although there is only one way in which a spherical s orbital ($l = 0$) can be oriented in space, p ($l = 1$) and d ($l = 2$) orbitals can point in different directions. There is only one s orbital per energy level, and its m_l value is 0. There are three different orientations for p orbitals and five orientations for d orbitals (see Figures 6.7 and 6.8). In fact, the number of orientations depends on the type of orbital. For a given value of l, there are $(2l + 1)$ integral values of m_l, as follows:

$$-l, (-l + 1), ...0, ...(+l - 1), +l$$

How many different orientations, or m_l values, are there when $l = 2$ (that is, how many different l orbitals)? If $l = 2$, then there are $[(2 \times 2) + 1]$, or five values of m_l, namely, $-2, -1, 0, 1,$ and 2. The number of m_l values indicates the number of orbitals in a subshell with a particular l value. In other words, there are five orbitals in the 3d subshell (and the 4d and 5d subshell, and so on). Because, according to Pauli, each orbital can hold two electrons, we need one last quantum number to distinguish their electronic addresses.

m_s

The **electron spin quantum number** (m_s) describes the spin of an electron in an orbital and can have values of only $+1/2$ and $-1/2$. Experiments on the emission spectra of hydrogen and sodium atoms indicated that lines in the emission spectra could be split by the application of an external magnetic field. Physicists explained these results by assuming that electrons act like tiny magnets. If electrons are thought of as spinning on their own axes, as Earth does, their magnetic properties can be accounted for. According to electromagnetic theory, a spinning charge generates a magnetic field, and it is this motion that causes an electron to behave like a magnet.

6.1 EXAMPLE

List the values of n, l, and m_l for orbitals in the 4d subshell.

Answer   As we saw earlier, the number given in the designation of the subshell is the principal quantum number, and d orbital corresponds to the angular momentum quantum number

$$n \searrow \quad l = 2 \swarrow$$
$$4d$$

The values of m_l can vary from $-l$ to l. Therefore, m_l can be $-2, -1, 0, 1,$ or 2 (which correspond to the five d orbitals).

6.2

EXAMPLE

What is the total number of orbitals associated with the principal quantum number n = 3?

Answer   *For n = 3, the possible values of l are 0, 1, and 2. Therefore, there is one 3s orbital (n = 3, l = 0, and m_l = 0); there are three 3p orbitals (n = 3, l = 1, and m_l = −1, 0, 1); there are five 3d orbitals (n = 3, l = 2, and m_l = −2, −1, 0, 1, 2). So the total number of orbitals is 1 + 3 + 5 = 9.*
   There's a short cut to this problem. The total number of orbitals is always given by n^2. You should check the validity of this formula by calculating the total number of orbitals for n = 1, 2, 3, and 4.

More About Orbital Shapes

Let's not forget that the quantum numbers represent addresses for electrons in orbitals. They are important because they tell us where the electrons are in an atom, and where an electron is in an atom is largely determined by the shape of its orbital. Strictly speaking, an orbital does not have a well-defined shape because the wave function characterizing the orbital extends from the nucleus to infinity. In that sense, it is difficult to say what an orbital looks like. On the other hand, it is certainly convenient to think of orbitals in terms of specific shapes, particularly in discussing the formation of chemical bonds between atoms.

   Although in principle an electron can be found anywhere, we know that most of the time it is quite close to the nucleus. For example, solving the Schrödinger equation for an electron in the 1s orbital gives roughly 90% probability of finding the electron within a sphere of radius 100 pm (1 pm = 1×10^{-12} m) surrounding the nucleus. Thus, we can represent the 1s orbital by drawing a *boundary surface diagram* that encloses about 90% of the total electron density in an orbital as shown in Figure 6.6. A 1s orbital represented in this manner is merely a sphere.

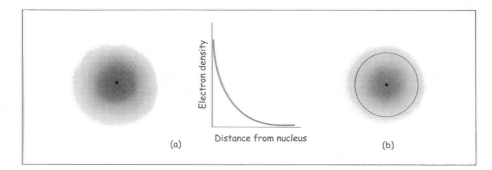

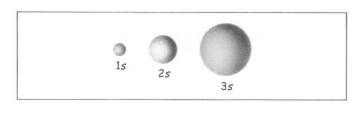

Figure 6.6 *Top panel:*
(a) Plot of electron density in the hydrogen 1s orbital as a function of the distance from the nucleus.
(b) Boundary surface diagram of the hydrogen 1s orbital.
Bottom panel: Boundary surface diagrams of the 1s, 2s, and 3s orbitals. Each sphere contains about 90% of the total electron density.

s (l = 0)

All *s* orbitals are spherical in shape but differ in size, which increases as the principal quantum number increases (that is, the size increases as follows: $1s < 2s < 3s$ ...). Although the details of electron density variation within each boundary surface are lost, there is no serious disadvantage. For us, the most important features of atomic orbitals are their shapes and *relative* sizes, which are adequately represented by boundary surface diagrams.

p (l = 1)

You should know that the *p* orbitals start with the principal quantum number $n = 2$. If $n = 2$, then the angular momentum quantum number l can have values of 0 and 1. As we saw earlier, when $l = 1$, the magnetic quantum number m_l can have values of $-1, 0, 1$. Starting with $n = 2$ and $l = 1$, we therefore have three $2p$ orbitals (Figure 6.7), which are oriented along the three axes: $2p_x$, $2p_y$, and $2p_z$. The letter subscripts indicate the axes along which the orbitals are oriented. These three *p* orbitals are identical in size, shape, and energy; they differ from one another only in orientation. Note, however, that there is no simple relation between the values of m_l and the *x*, *y*, and *z* directions. For our purpose, you need only remember that because there are three possible values of m_l, there are three *p* orbitals with different orientations.

The boundary surface diagrams of *p* orbitals show that each *p* orbital can be thought of as two lobes; the nucleus is at the center of the *p* orbital. Like *s* orbitals, *p* orbitals increase in size from $2p$ to $3p$ to $4p$ orbital, and so on.

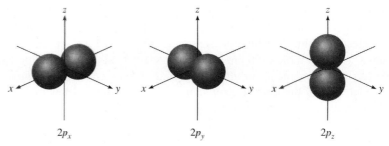

$2p_x$ $2p_y$ $2p_z$

Figure 6.7 The shapes of *p* orbitals

d (l = 2)

You should know that the *d* orbitals start with the principal quantum number $n = 3$. If $n = 3$, then the angular momentum quantum number l has values of 0, 1, and 2. When $l = 2$, there are five values of m_l ($-2, -1, 0, 1, 2$), which correspond to five *d* orbitals: $3d_{xy}$, $3d_{yz}$, $3d_{xz}$, $3d_{x^2-y^2}$, and $3d_{z^2}$ (Figure 6.8). As in the case of the *p* orbitals, the different orientations of the *d* orbitals correspond to the different values of m_l, but again there is no direct correspondence between a given orientation and a particular m_l value. All the $3d$ orbitals in an atom are identical in energy. The *d* orbitals for which *n* is greater than 3 ($4d$, $5d$, . . .) have similar shapes.

Although there are other types of orbitals, the *s*, *p*, and *d* orbitals are the only ones you need to know about the shapes and orientations. Try not to be intimidated by the strange shapes of orbitals, especially the *d* orbitals.

You have seen that quantum numbers are important for determining the sizes (and energies) of orbitals, their shapes, as well as the number of orbitals in a particular shell. Our next step is to see how the electrons are distributed among the various orbitals for

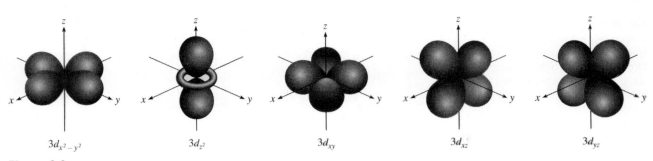

$3d_{x^2-y^2}$ $3d_{z^2}$ $3d_{xy}$ $3d_{xz}$ $3d_{yz}$

Figure 6.8 The shapes of *d* orbitals

a given atom. Our earlier analogy of the atomic hotel shows that each orbital, or room, can accommodate a total of 2 electrons. Therefore, we know that 2 electrons can be housed on the first floor (the 1*s* orbital), 8 electrons on the second (the 2*s* and 2*p* orbitals), and 18 on the third (that is, 2 in the 3*s* orbital, 6 in the 3*p* orbitals, and 10 in the 3*d* orbitals). We can use quantum numbers to describe the location of the electrons in an atom just like we can list the addresses of the hotel occupants. But describing an atom with a lot of electrons using a list of the four quantum numbers for each electron would be unnecessarily cumbersome. There is an easier way. . . .

Which Orbitals Are Occupied in Atoms?

Now that you know about the types of atomic orbitals, we are ready to discuss how they are filled in atoms. Let's consider the hydrogen atom first. Which orbital do you think will be occupied by the lone electron? You know that this electron would want to be in the lowest-energy state, or level, if it isn't excited by an external source. Because there is only one orbital, 1*s,* in the lowest level, this is where you will find the electron. A hydrogen atom with its electron in the 1*s* orbital is called the ground-state H atom, because the energy of the electron is at its minimum. For atoms having more than one electron, called *many-electron atoms,* we can again obtain the ground-state arrangement if we place the electrons in the order of increasing energy, that is, from the lowest-energy level up to higher ones. Referring again to our atomic hotel analogy, the hotel manager fills the rooms on the ground floor first and then fills the rooms on the second, third . . . floors in that order. An atom with its electrons filled in this way will be in its most stable state because the total energy of the electrons is at its minimum.

It should make sense then that an electron will not occupy the fourth shell, for example, if there are empty orbitals in the third shell. But, which orbital would have an electron if the shell contains different types of orbitals? The one with the lowest energy, of course. So we need to know the relative energies of the different types of orbitals. For a many-electron atom, the order within a given shell is exactly what you might expect: *s* < *p* < *d,* and so on. One way to think about this order is to consider that the average distance of the electrons increases as you go from *s* to *p* to *d,* and so forth. So in the third shell, the 3*s* orbital has lower energy than each of the three 3*p* orbitals, which have lower energy than each of the five 3*d* orbitals. Let's consider the ground-state sodium atom with 11 electrons. To identify the locations of the electrons, we can list the atomic orbitals in order of increasing energy and add two electrons to each orbital, starting with the lowest-energy 1*s* orbital and filling them until all 11 are distributed. So there would be 2 electrons in the 1*s* orbital, 2 electrons in the 2*s* orbital, 6 electrons in the three 2*p* orbitals, and 1 electron in the 3*s* orbital—that's 11 in all (Figure 6.9).

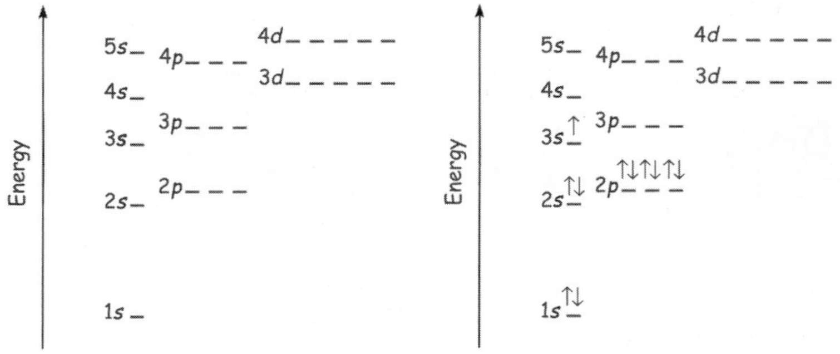

Figure 6.9 *Left:* Orbital energy levels for a many-electron atom. *Right:* Occupation of orbitals in a sodium atom. Electrons are indicated as arrows pointing up or down, corresponding to the opposite spins they have when paired in an orbital.

Electron Configuration | *The atomic census*

For convenience, chemists use a shorthand method, called **electron configuration,** for describing the location of electrons in atoms and ions. The electron configuration is a list of all the occupied orbitals with a superscript after each indicating the number of electrons. For hydrogen, the ground-state electron configuration is $1s^1$, and for sodium, it's $1s^2 2s^2 2p^6 3s^1$. Isn't this a lot easier than writing out all of the quantum numbers? As you move to larger atoms, it gets a little more complicated, but the easiest way to figure out the electron configuration of an atom is to use the periodic table.

The Periodic Table Is Based on Electron Configurations

Although Bohr is best known for his model of the hydrogen atom, what he was especially interested in was the relationship between the periodic table and the arrangement of electrons in atoms. Bohr suggested that the chemical and physical properties of atoms depend on the way electrons are arranged around the nucleus. Long before the discovery of electrons or the nucleus, Mendeleev had already arranged the elements according to their chemical properties and molar masses (as you saw in Chapter 2). So Bohr thought there must be a connection between the arrangement of electrons in atoms and the periodic table. It turns out that the elements in the same group on the periodic table have the same electron configuration in their outermost shell. In fact, it was Bohr who first used the term shell to describe the energy levels of an atom. He proposed that each shell can accommodate a certain number of electrons, and the degree to which a shell is filled determines an atom's chemical properties; a filled shell gives chemical stability. According to Bohr, the stable noble gases should have fully filled shells.

Let's explore Bohr's idea of full shells by considering the number of electrons in the noble gases in the context of what we know about the number of subshells. The number of electrons in He, Ne, Ar, and Kr are 2, 10, 18, and 36, respectively. If these atoms have all their shells filled, we would infer that the first shell holds 2 electrons; the second, 8; the third, 8; and the fourth, 18. Although the numbers 2, 8, and 18 are consistent with the number of electrons predicted by quantum mechanics, we would expect argon to have 36 electrons. That is, there should be 2 in the first shell (in the 1s), 8 in the second shell (in the 2s and 2p's), and 18 in the third shell (in the 3s, 3p's, and 3d's).

A likely explanation for argon having only 8 in its third shell instead of 18 is that its 3s and 3p orbitals are filled, but not the 3d. Apparently krypton, with 18 electrons more than argon, has a filled d subshell in addition to the filled 4s and 4p subshell. However, this filled d subshell must be the 3d subshell because quantum mechanics tells us that the 3d is the lowest-energy d subshell available. This doesn't

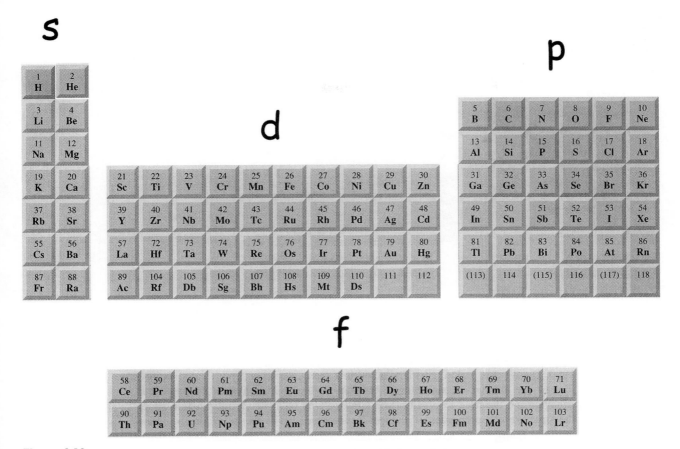

Figure 6.10 All of the elements in each section of the periodic table have their outermost electrons in the same type of orbital.

present any problems in terms of the order in which orbitals are filled until you consider the potassium atom with 19 electrons. Potassium is in the same group (1A) as Li and Na—all having one electron in an outermost s orbital; Li is $1s^2 2s^1$, and Na is $1s^2 2s^2 2p^6 3s^1$. It follows that the chemically similar K would have an electron configuration of $1s^2 2s^2 2p^6 3s^2 3p^6 4s^1$. The outermost electron is in the $4s$ orbital, and the $3d$ subshell is empty! This changes our rules regarding the order of filling orbitals and this isn't the only case. For example, the $6s$ orbital is filled before the $5d$ and the $4f$ orbitals. The good news is that you don't have to memorize a complicated filling order because it's all laid out in the periodic table.

Let's start by breaking the periodic table into four parts (s, p, d, f) as shown in Figure 6.10. Note that the number of elements in each horizontal row is 2, 6, 10, 14. That's exactly the number of available spots in the subshells $s, p, d,$ and f. It turns out that Groups 1A and 2A atoms have an s as their outer subshell, Groups 3A–8A have p as their outer subshell, the transition metals have a d as their outer subshell, and the inner transition metals ($Z = 58-71$ and $90-103$) have an f as their outer subshell. Moreover, atoms in the same groups have the same number of electrons in their outer subshell (as Bohr suggested), and that number increases across the periodic table. For example, the outer subshell configurations for the second period elements are $2s^1, 2s^2, 2p^1, 2p^2, 2p^3, 2p^4, 2p^5,$ and $2p^6$. So the periodic table is a map of electron configurations as shown in Figure 6.11.

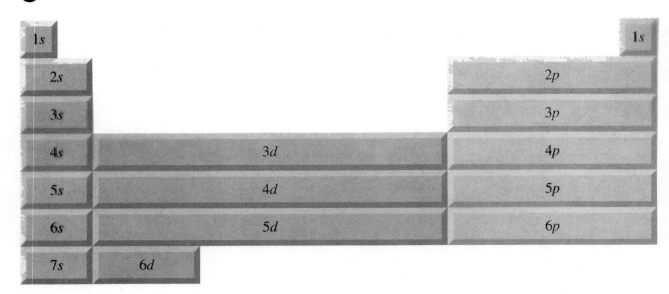

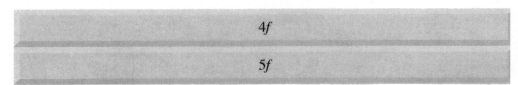

Figure 6.11   Periodic guide for writing electron configurations

EXAMPLE

Write the electron configuration of sulfur (Z = 16).

Answer   *Sulfur has 16 electrons. It takes 10 electrons to complete the first and second periods ($1s^2 2s^2 2p^6$). This leaves 6 electrons to fill the 3s orbital and partially fill the 3p orbitals. Thus, the electron configuration of S is*

$$1s^2 2s^2 2p^6 3s^2 3p^4$$

(Check the website for more problems on writing electron configurations.)

Electron Configurations of Ions

In addition to writing ground-state electron configurations of elements, you should also learn to do the same for ions. There is an easy way to do this for representative elements. When the atom of a representative element is converted to an anion, it wants to look just like the noble gas immediately following it. For example, when fluorine is converted to fluoride (F^-), its electron configuration becomes the same as that for Ne ($1s^2 2s^2 2p^6$). This is also true for O^{2-} and N^{3-}. The ions F^-, O^{2-}, and N^{3-} and the atom Ne are said to be *isoelectronic* because they have the same (*iso*) number of electrons and therefore must have the same ground-state electron configuration. On the other hand, when the atom of a representative element is converted to a cation, it becomes isoelectronic with the noble gas immediately preceding it. Examples are Na^+, Mg^{2+}, and Al^{3+} ($1s^2 2s^2 2p^6$). Again, these ions and Ne all have the same ground-state electron configuration.

The situation is more involved for transition metals. We will just focus on the first row transition metals (Sc to Cu). When such a metal forms a cation, it does not become isoelectronic with the noble gas preceding it. And it can form different cations like Fe^{2+} and Fe^{3+}. Furthermore, although the order of filling electrons is from $4s$ to $3d$, the order of removing electrons from the atom is not the reverse ($3d \longrightarrow 4s$), but instead, the $4s$ electrons are removed before the $3d$ electrons. For example, Mn is $[Ar]4s^2 3d^5$, and Mn^{2+} is not $[Ar]4s^2 3d^3$ but $[Ar]3d^5$. (Note that [Ar] is called the argon core, which is a shorthand notation for the electron configuration of argon.) Remember this when you write the electron configuration of a transition metal cation: The $4s$ electrons are removed before the $3d$ electrons in forming a cation.

How Are the Electrons Arranged in Partially Filled Subshells?

Electron configurations tell us a lot about the number of electrons in the different orbitals (subshells) of an atom. But, they don't tell us anything about how the electrons are distributed in the subshells. For example, consider the nitrogen atom with seven electrons. Its electron configuration is $1s^2 2s^2 2p^3$. The p subshell is not filled—it has three vacancies. Where are those vacancies? Where are the three electrons, and what are their spins? There are several possibilities. There could be two in any of the three p orbitals and the third in either of the remaining orbitals, or there could be one in each orbital. The number of possibilities grows when you consider the three electrons could have any combination of spins. It turns out there is one electron in each $2p$ orbital and all three electrons have the same spin. This mode of filling orbitals follows a general rule that applies to all atoms: **The most stable arrangement of electrons in subshells is the one with the greatest number of parallel spins.** This rule was proposed by the German physicist *Frederick Hund* and is appropriately named **Hund's rule.** Hund's rule dictates that all three $2p$ electrons have spins parallel to one another.

Hund's rule can be tested experimentally because spinning electrons are like little magnets. The test is simple: see if an atom is attracted to a magnet. To understand the test, you have to know that a spinning electron generates a magnetic field and that a pair of electrons with opposite spins cancel each other's magnetic field; unpaired electrons do not. As a result, atoms with all their electrons paired are not attracted to a magnet, and they are said to be **diamagnetic.** Atoms with unpaired electrons are attracted to a magnet and are called **paramagnetic.** Now you might say that this doesn't prove that nitrogen has three unpaired electrons; because there is an odd number of electrons, nitrogen would be paramagnetic no matter how the $2p$ orbitals are filled. You would be right, but the amount of attraction can be quantified. Consider boron ($1s^2 2s^2 2p^1$) and carbon ($1s^2 2s^2 2p^2$) with one and two electrons in their $2p$ subshell, respectively. Both are paramagnetic, but carbon is attracted to a magnet more strongly than boron, and nitrogen is attracted more than carbon. So in a quantitative way, we can establish the number of unpaired spins in an atom experimentally. The fact that some atoms have magnetic properties also supports the Pauli exclusion principle. Can you explain why?

Diagramming Orbitals

What if we wanted to give a representation of all the electrons in an atom, including the distribution of electrons in individual orbitals? Drawing the orbitals for atoms with a lot of electrons would be very messy. Chemists have devised a very simple depiction called the **orbital diagram.** In an orbital diagram, the orbitals are drawn as boxes arranged vertically so that those with higher energy are above those with lower energy,

Figure 6.12 Orbital diagrams for iron and bromine

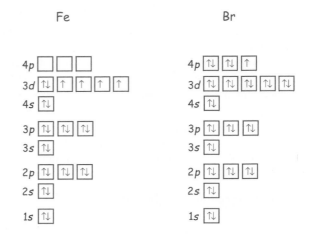

and the electrons are drawn as arrows pointing up or down to indicate the opposite spins. Consider the examples in Figure 6.12. We've indicated the unpaired electrons all pointing up, but you could have them all point down. It doesn't matter as long as they are all parallel to one another according to Hund's rule.

The Periodic Table Is a Tool for Predicting Atomic Properties

The number of electrons in an atom's outermost shell (or **valence shell**) tells us a lot about its properties and behavior. For example, the alkali metals (Group 1A) will readily lose its single electron in the outermost shell (ns^1) to become a $+1$ cation and have an electron configuration like a noble gas preceding it. On the other hand, the halogens want to gain an electron to be like a noble gas. **The basic idea is this: Atoms of representative elements react chemically to have an electron configuration like a noble gas.**

We have already seen in a very broad sense how the periodic table can be used to predict chemical and physical properties. For example, we know that everything to the left of the zigzag line behaves like a metal and everything to the right behaves like a nonmetal. As we mentioned in Chapter 2, there is not an abrupt change in properties at the line, but rather a gradual decrease in metallic properties (and a corresponding increase in nonmetallic properties) as you move across and up the periodic table.

As you already know, the metals and nonmetals can best be distinguished chemically by whether they like to lose or gain electrons. Metals like to lose electrons, and nonmetals like to gain electrons. It turns out that the ability to lose or gain electrons can be quantified. The quantitative measure of the abiltiy to lose electrons is called **ionization energy,** and the quantitative measure of the ability to gain electrons is called **electron affinity.** Not surprisingly, there is a trend in these two properties as you move across and down the periodic table. But in order to understand these trends, you must first learn how the size of an atom relates to its position on the periodic table. The size of an atom, measured by its radius, is an important factor in assessing the trends in ionization energy and electron affinity.

Atomic Radius

The size of an atom, measured either by its volume or simply the radius, influences properties such as density and melting point, as well as shape of molecules. The **atomic radius** measured in metals is one-half the distance between the centers of two adjacent atoms; for nonmetals, it's one-half the distance between centers of atoms in a diatomic molecule (Figure 6.13). It's useful for our purpose here to think about the radius as the distance between the nucleus and the outermost electron. Generally, the size of an atom decreases as we move from left to right in a particular period. This is not what you would expect because atoms get heavier as you move across a period, but let's examine this trend a little more closely.

To understand the trend in atomic radius across a period, we have to consider what actually determines the size of an atom. As you already know, the negatively charged electrons are attracted to the positive nucleus; this attraction is the primary basis for an atom's size. The more positive the nucleus, the stronger the pull on the electrons, and the smaller the radius. This attraction can be attenuated by a couple of factors. First, completely filled inner shells will act like a shield to reduce the electrostatic attraction between protons in the nucleus and the electrons in the outer shells. In other words, the pull of the nucleus on electrons in, say, the 3s orbital, is less if the first and second shells are filled with electrons than if there were no electrons in the inner shells. In this hypothetical case, the atom with the filled shells would be larger because the electrons are not pulled in as strongly. This effect also occurs to a lesser degree when subshells are filled. That is, a filled subshell will reduce somewhat the electrostatic attraction between the nucleus and the electrons in higher-energy subshells. For example, a filled 2s orbital has a small shielding effect on the electrons in the 2p orbitals. The electrostatic pull by the nucleus is also reduced slightly by the repulsion from electrons in the same subshell. But the overriding effect is the number of protons in the nucleus.

Let's consider the elements in the second period. Starting with lithium, we have three protons in the nucleus and three electrons—two in the 1s orbital and one in the 2s orbital. The electron in the 2s orbital largely determines the size of lithium. That is, its distance from the nucleus is the atomic radius. Now let's consider beryllium with four protons and four electrons. Here we have two electrons in the 2s orbital feeling a pull of four protons. Because the fourth electron is added to the same shell (and the same subshell), there is no additional shielding of the nuclear charge, so both electrons in the 2s orbital are feeling the pull of +4 from the nucleus (shielded in part by the filled 1s orbital). By contrast, the 2s electron in lithium feels +3 (75% of the pull felt by the beryllium atoms). Therefore, the lithium electron will be farther away from the nucleus than the 2s electrons in beryllium and will have a larger radius. As we continue to move across the second period, we are adding more positive charge to the nucleus, while electrons are added to the same shell. More nuclear pull without any additional shell shielding means the atomic radius gets smaller as we move across the period. Moving down a particular group, we see the atomic size increases. This is not surprising because the principal quantum number *n* increases and the outer shell continues to expand. The sizes of the representative elements and the general trends are shown in Figure 6.14.

Ionic Radius

Ionic radius is the radius of a cation or an anion. When an atom is converted to an ion, we expect the size of the ion to be different from that of the parent atom. If the atom is converted to a cation, the cation will be smaller than the atom because the nuclear charge remains the same but there is now a decrease in electron-electron repulsion. On the other hand, an anion will be larger than the parent atom because the extra electron(s) will expand the domain of the electron density. Figure 6.15 shows the ionic radii of a number of ions. Moving down a group, the size of the ions increases, which is analogous to the trends observed for atoms. No clear trends exist as we move across a period because both cations and anions are involved.

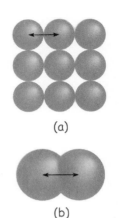

Figure 6.13 Atomic radius in (a) metals and (b) a diatomic molecule

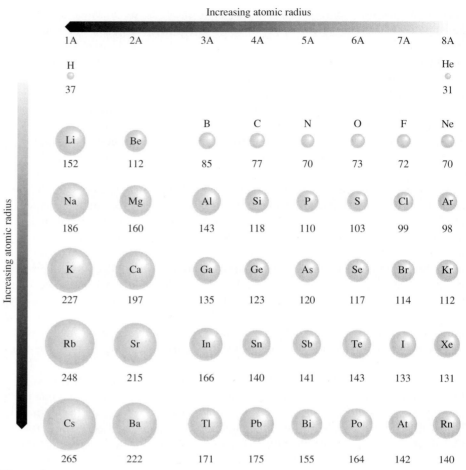

Figure 6.14 Atomic radii (in picometers) of representative elements

Ionization Energy

Chemical properties of any atom are determined by the configuration of the atom's outer electrons, called **valence electrons.** This makes sense because when atoms approach each other, it is these electrons that feel each other's presence. The stability of these outermost electrons is reflected directly in the atom's *ionization energy.* Ionization energy is the minimum energy required to remove an electron from a gaseous atom in its ground state. The magnitude of ionization energy is a measure of the effort required to force an atom to give up an electron, or of how "tightly" the electron is held in the atom. The higher the ionization energy, the more tightly held is the electron and the more difficult it is to remove the electron.

For a many-electron atom, the amount of energy required to remove the first electron from the atom in its ground state

$$\text{energy} + \text{X}(g) \longrightarrow \text{X}^{+}(g) + e^{-} \qquad\qquad I_1$$

is called the *first ionization energy* (I_1). In the preceding equation, X represents a gaseous atom of any element, and e^{-} is an electron. The *second ionization energy* (I_2) and the *third ionization energy* (I_3) are shown in the following equations:

$$\text{energy} + \text{X}^{+}(g) \longrightarrow \text{X}^{2+}(g) + e^{-} \qquad\qquad I_2$$

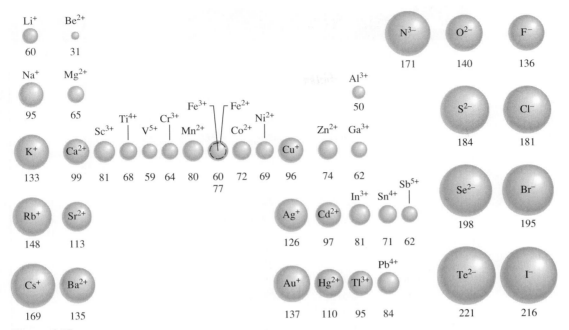

Figure 6.15 Radii of ions (in picometers) of familiar elements

$$\text{energy} + X^{2+}(g) \longrightarrow X^{3+}(g) + e^- \qquad\qquad I_3$$

The pattern continues for the removal of subsequent electrons.

When an electron is removed from a neutral atom, the repulsion among the remaining electrons decreases, and the atomic radius also decreases. Because the nuclear charge remains constant, more energy is needed to remove another electron from the positively charged ion. Thus, for the same element, ionization energies always increase in the following order:

$$I_1 < I_2 < I_3 < \ldots$$

Table 6.1 lists the ionization energies of a number of elements.

By convention, energy absorbed by atoms (or ions) in the ionization process has a positive value. Thus, ionization energies are all positive quantities. This should make sense when you realize that the energy absorbed is measured as the enthalpy change, ΔH. Because the atomic radius decreases across a period, the ionization energy increases from left to right due the increasing nuclear attraction for the electrons; that is, the electrons are held tighter.

The Group 1A elements (the alkali metals) have the lowest ionization energies. Each of these metals has one valence electron (the outermost electron configuration is ns^1) that is effectively shielded by the completely filled inner shells. Consequently, it is energetically easy to remove an electron from the atom of an alkali metal to form a unipositive ion (Li^+, Na^+, K^+, . . .).

The Group 2A elements (the alkaline earth metals) have higher first ionization energies than the alkali metals do. The alkaline earth metals have two valence electrons (the outermost electron configuration is ns^2). Because these two s electrons do not shield each other well, the effective nuclear charge (that is, the charge felt by an electron) for an alkaline earth metal atom is larger than that for the preceding alkali

Table 6.1 Ionization Energies (kJ/mol) of the First 20 Elements

Z	Element	First	Second	Third	Fourth	Fifth	Sixth
1	H	1,312					
2	He	2,373	5,251				
3	Li	520	7,300	11,815			
4	Be	899	1,757	14,850	21,005		
5	B	801	2,430	3,660	25,000	32,820	
6	C	1,086	2,350	4,620	6,220	38,000	47,261
7	N	1,400	2,860	4,580	7,500	9,400	53,000
8	O	1,314	3,390	5,300	7,470	11,000	13,000
9	F	1,680	3,370	6,050	8,400	11,000	15,200
10	Ne	2,080	3,950	6,120	9,370	12,200	15,000
11	Na	495.9	4,560	6,900	9,540	13,400	16,600
12	Mg	738.1	1,450	7,730	10,500	13,600	18,000
13	Al	577.9	1,820	2,750	11,600	14,800	18,400
14	Si	786.3	1,580	3,230	4,360	16,000	20,000
15	P	1,012	1,904	2,910	4,960	6,240	21,000
16	S	999.5	2,250	3,360	4,660	6,990	8,500
17	Cl	1,251	2,297	3,820	5,160	6,540	9,300
18	Ar	1,521	2,666	3,900	5,770	7,240	8,800
19	K	418.7	3,052	4,410	5,900	8,000	9,600
20	Ca	589.5	1,145	4,900	6,500	8,100	11,000

metal atom. Alkaline earth compounds contain dipositive ions (Mg^{2+}, Ca^{2+}, Sr^{2+}, Ba^{2+}), which are isoelectronic with the unipositive alkali metal ions preceding them in the same period (and they are all isoelectronic with the noble gases preceding them).

From the preceding discussion, we see that metals have relatively low ionization energies, whereas nonmetals possess much higher ionization energies. The ionization energies of the metalloids usually fall between those of metals and nonmetals. The difference in ionization energies suggests why metals always form cations and nonmetals form anions in ionic compounds. For a given group, the ionization energy decreases with increasing atomic number (that is, as we move down the group). Elements in the same group have similar outer electron configurations. However, as the principal quantum number n increases, so does the average distance of a valence electron from the nucleus. A greater separation between the electron and the nucleus means a weaker attraction, so the electron becomes increasingly easier to remove as we go from element to element down a group. Thus, the metallic character of the elements within a group increases from top to bottom. This trend is particularly noticeable for elements in Groups 3A to 7A. For example, in Group 4A we note that carbon is a nonmetal, silicon and germanium are metalloids, and tin and lead are metals.

Electron Affinity

Atoms can also gain one or more electrons to become anions, and the tendency for them to do so is measured by their *electron affinity*. Because metals tend to form cations, we expect them to have little or no affinity for electrons; that is, they should have a small electron affinity. Atoms of nonmetallic elements like N, O, F, and Cl, on the other hand, can readily form anions (N^{3-}, O^{2-}, F^-, and Cl^-), so these elements have a high electron affinity.

Table 6.2	Electron Affinities (kJ/mol) of Representative Elements						
1A	2A	3A	4A	5A	6A	7A	8A
H 73							He < 0
Li 60	Be ≤ 0	B 27	C 122	N 0	O 141	F 328	Ne < 0
Na 53	Mg ≤ 0	Al 44	Si 134	P 72	S 200	Cl 349	Ar < 0
K 48	Ca 2.4	Ga 29	Ge 118	As 77	Se 195	Br 325	Kr < 0
Rb 47	Sr 4.7	In 29	Sn 121	Sb 101	Te 190	I 295	Xe < 0
Cs 45	Ba 14	Tl 30	Pb 110	Bi 110	Po ?	At ?	Rn < 0

Consider the process in which a gaseous fluorine atom accepts an electron:

$$F(g) + e^- \longrightarrow F^-(g) \qquad \Delta H = -328 \text{ kJ/mol}$$

This is an exothermic process, as indicated by the negative sign for the enthalpy change. The electron affinity of flourine, however, is assigned a positive value of +328 kJ/mol. Table 6.2 lists the electron affinity values of a number of elements. Note that the more positive the electron affinity, the greater is the tendency for the atom of the element to accept an electron.

Now the formation of an ionic compound like NaCl begins to make sense. Sodium has a low ionization energy (readily gives up its $3s^1$ electron) and a low electron affinity (little tendency to gain an electron), and chlorine has a high ionization energy (its $3p$ electrons are tightly held) and a high electron affinity (wants to take up an electron to fill its $3p$ subshell). If we heat sodium metal in chlorine gas, the product NaCl contains Na$^+$ and Cl$^-$ ions:

$$2Na(l) + Cl_2(g) \longrightarrow 2NaCl(s)$$

See website for a discussion of the energetics of the formation of ionic compounds.

Test your understanding of the material in this chapter

IMPORTANT TERMS

Explain the following terms in your own words:

angular momentum quantum number, p. 89
atomic radius, p. 98
diamagnetic, p. 97
electron affinity, p. 98
electron configuration, p. 94
electron spin quantum number, p. 90
excited level, p. 85
excited state, p. 85
ground level, p. 85
ground state, p. 85
Hund's rule, p. 97
ionic radius, p. 99
ionization energy, p. 98
isoelectronic, p. 96
magnetic quantum number, p. 90
orbital, p. 87
orbital diagram, p. 97
paramagnetic, p. 97
Pauli exclusion principle, p. 88
principal quantum number, p. 89
quanta, p. 82
quantum mechanics, p. 87
quantum numbers, p. 88
quantum theory, p. 82
valence electron, p. 100
valence shell, p. 98

7

Chemical Bonding
The glue that holds atoms together

n Chapter 6 you learned that atomic properties such as size and the ability to lose or gain electrons are determined by the way electrons are arranged in atoms. In this chapter you will see how knowing the arrangement of electrons in atoms enables us to understand how and why atoms of different elements react to form chemical bonds. We will focus mostly on covalent bonds, the attractive forces that hold atoms together in molecules. As a starting point, you can think of the covalent bond as a kind of "chemical glue" involving a pair of electrons shared by two atoms.

Of course, electrons are not glued together, so this analogy doesn't really help us understand anything more than the fact that atoms are joined together. It doesn't help us predict anything about the behavior of molecules. Knowing how atoms actually form covalent bonds is very important because it helps us predict the way atoms are arranged in molecules and how the molecules fill space. If we can visualize this arrangement, we can understand the shape or geometry of molecules. Just as knowing the shape of a macroscopic object can tell us how the object will behave (e.g., a spherical ball will roll if given a push), knowing the shape of molecules suggests how they will behave. Throughout this book, we have encouraged you to imagine atoms, molecules, and ions as objects that have well-defined shapes. Understanding how bonding determines the three-dimensional shapes of molecules is one of the most important keys to understanding chemistry because it allows us to visualize the virtually invisible world of molecules. In this chapter we will study how atoms are joined together to form a handful of basic geometric shapes, and we will describe the way these shapes help us predict some fundamental molecular properties like solubility and the forces that hold molecules together in liquid and solid states. Figure 7.1 shows the structures of some molecules whose shapes give rise to very complex properties. One of the most empowering aspects of learning chemistry is that you can understand these complex properties at a fundamental level by knowing a few basic bonding concepts we will describe in this chapter.

Electronic hand-holding links atoms in a stable union.

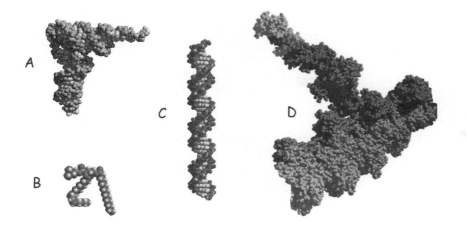

Figure 7.1 Structures of some cool molecules.
(a) The structure of transfer RNA (tRNA) provides for reading the genetic code on one end and making proteins on the other end.
(b) The kinked structure of phosphatidyl choline maintains the fluidity of cell membranes.
(c) The structure of DNA provides for a copying mechanism.
(d) The structure of the proteins actin and myosin provide for muscle contraction.

Before we begin our study of chemical bonding, let's first review what you already know:

1. Atoms, ions, and molecules react by transferring or sharing electrons (p. 15).
2. Atoms react to achieve electron configurations like noble gases in their valence shells (p. 98).
3. Metals readily lose electrons to electron-hungry nonmetals; the resultant cations and anions are attracted to each other by ionic bonds in ionic compounds (p. 18).
4. Nonmetals share electrons with other nonmetals to form covalent bonds in molecules (p. 16).
5. Electron affinity, the attraction an atom has for electrons outside the atom, increases as you move across the periods and decreases as you move down groups on the periodic table.
6. Electrons in a bond between atoms with different electron affinities are not shared equally, and this causes some molecules to be polar (p. 103).

So you already know all of the basic features of bonding, but not quite enough to really understand how and why it happens.

The Nature of Chemical Bonding

To begin to understand the nature of bonding, let's think about the implications for bonding suggested by the periodic trends in electron affinity. Because there is a gradual increase in electron affinity across a particular period, the extent of electron sharing between atoms varies quite a bit. For this reason, one can think of chemical bonding as a continuum, ranging from the purely covalent bond in which the electrons are equally shared, as in H_2, to the ionic bond between atoms with vastly different electron affinities, like CsF (with atoms from opposite ends of the periodic table). Somewhere between these extreme cases are **polar molecules,** like HCl, where the bond is said to be *polarized*. In H_2, the electrons from the two atoms are shared equally because both atoms have exactly the same affinity for electrons. In HCl, the chlorine atom, having a greater attraction for electrons, pulls the two shared electrons (its own and the one from hydrogen) toward itself. Such a bond is called a polar covalent bond. In CsF, the electron from cesium is completely transferred to the fluorine atom, yielding a pair of oppositely charged ions that attract one another just like opposite poles of a magnet.

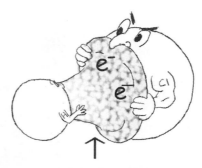

Polarized covalent
bond in HCl molecule

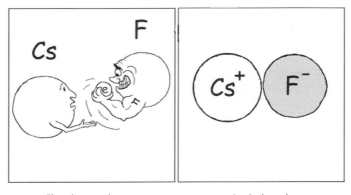

Fluorine grabs an
electron from cesium

Ionic bond

Can we quantify the difference in electron affinity in bonded atoms? If we wanted to predict the extent to which a bond is polarized (or to what extent the shared electrons are pulled toward one atom in a chemical bond), it would be useful to have a way to quantify the attraction an atom has for electrons in a bond.

Electronegativity: The Appetite for Electrons

In 1939 the great American chemist *Linus Pauling* introduced a property called **electronegativity** as a quantitative measure of an atom's tendency to attract electrons in a chemical bond. The concept of electronegativity is extremely important in understanding the chemical and physical properties of compounds. It determines the types of bonds that are formed—covalent, polar covalent, or ionic—and it helps us predict the way molecules will react. Like electron affinity, which is the energy change when an isolated atom grabs an electron to become an anion, electronegativity is also a measure of an atom's attraction for electrons, but it pertains to the attraction within a molecule. On the electronegativity scale shown in Figure 7.2, fluorine has an electronegativity of 4.0, making it the most electronegative element. Oxygen (3.5), nitrogen (3.0), and chlorine (3.0) are also highly electronegative. The electronegativities of the alkali metals and most of the alkaline earth metals are around 1.0 or less.

The larger the electronegativity of an element, the greater the tendency of an atom of that element to pull electrons toward itself in a bond. Earlier in this book, we used

Increasing electronegativity

Increasing electronegativity

1A																	8A
H 2.1	2A											3A	4A	5A	6A	7A	
Li 1.0	**Be** 1.5											**B** 2.0	**C** 2.5	**N** 3.0	**O** 3.5	**F** 4.0	
Na 0.9	**Mg** 1.2	3B	4B	5B	6B	7B	⌐—8B—⌐			1B	2B	**Al** 1.5	**Si** 1.8	**P** 2.1	**S** 2.5	**Cl** 3.0	
K 0.8	**Ca** 1.0	**Sc** 1.3	**Ti** 1.5	**V** 1.6	**Cr** 1.6	**Mn** 1.5	**Fe** 1.8	**Co** 1.9	**Ni** 1.9	**Cu** 1.9	**Zn** 1.6	**Ga** 1.6	**Ge** 1.8	**As** 2.0	**Se** 2.4	**Br** 2.8	
Rb 0.8	**Sr** 1.0	**Y** 1.2	**Zr** 1.4	**Nb** 1.6	**Mo** 1.8	**Tc** 1.9	**Ru** 2.2	**Rh** 2.2	**Pd** 2.2	**Ag** 1.9	**Cd** 1.7	**In** 1.7	**Sn** 1.8	**Sb** 1.9	**Te** 2.1	**I** 2.5	
Cs 0.7	**Ba** 0.9	**La-Lu** 1.0-1.2	**Hf** 1.3	**Ta** 1.5	**W** 1.7	**Re** 1.9	**Os** 2.2	**Ir** 2.2	**Pt** 2.2	**Au** 2.4	**Hg** 1.9	**Tl** 1.8	**Pb** 1.9	**Bi** 1.9	**Po** 2.0	**At** 2.2	
Fr 0.7	**Ra** 0.9																

Figure 7.2 Electronegativities of common elements

ELECTRONEGATIVITY IS A MEASURE OF AN ATOM'S ATTRACTION FOR ELECTRONS IN A CHEMICAL BOND.

In addition to his pioneering work on the nature of chemical bonding, Pauling revolutionized the study of chemistry by building models of molecules.

the analogy of children playing with toys to describe the sharing of electrons. A big kid will take toys away from a little kid, but two big kids will agree to share the toys because they can't take them away from each other. The nonmetals are the big kids, the ones with the high electronegativities, while the metals, with the low electronegativities, are the little kids.

The Ionic Bond | *Positive attracts negative*

When atoms of the elements with large differences in electronegativity are brought together, ionic compounds are usually formed:

Figure 7.3 Na$^+$ and Cl$^-$ ions in the solid state shown as a single layer (*left*) and as they would be arranged in three dimensions (*right*)

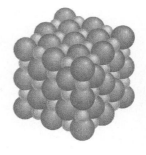

$$2Na(s) + Cl_2(g) \longrightarrow 2NaCl(s)$$

$$2Ca(s) + O_2(g) \longrightarrow 2CaO(s)$$

In an ionic compound, the forces that hold the ions together are the strong electrostatic forces between the cations and anions. For this reason, ionic compounds are always solids at room temperature, and their melting points are in the high hundreds and even over a thousand degrees Celsius. To maximize net attraction in solid NaCl, each cation is surrounded by six Cl$^-$ ions, and vice versa, as shown in Figure 7.3. Similar arrangements are found for other ionic compounds. Of course, an ionic compound may contain polyatomic cations or polyatomic anions such as CaSO$_4$ and NH$_4$Cl. In all cases, the structure of the compound is simply a basic unit of oppositely charged ions repeated indefinitely in a three-dimensional array.

The Covalent Bond | *The glue between atoms*

In 1916 the American chemist **Gilbert Lewis** introduced the concept of the covalent bond to explain the stability of molecules. He defined the covalent bond as the sharing of electrons between atoms. Lewis suggested that atoms share electrons in order to have a more stable electron configuration, namely that of a noble gas. The electrons involved in forming covalent bonds are valence electrons, those in the outermost shell of the atom.

Lewis first sketched his idea about the octet rule on the back of an envelope.

In Chapter 2 we told you to remember that carbon forms four bonds, nitrogen forms three bonds, oxygen forms two bonds, and hydrogen forms only one bond. Now that you know something about electron configurations and the way electrons are paired in orbitals, it will be easy to understand why atoms form a specific number of bonds. Lewis introduced a convenient way to represent valence electrons in atoms—as dots around the four sides of the chemical symbol for the element, as shown in Figure 7.4. Inspection of the electron dot symbols suggests that for every atom except hydrogen, a specific number of electrons must pair up in order to have the same number of valence electrons as a noble gas, that is, eight.

Figure 7.4 Electron dot symbols for some common nonmetals

For example, two hydrogen atoms readily react to form a hydrogen molecule where each hydrogen feels like it has two electrons, the same as helium; each dot denotes the lone electron on the H atom.

$$H\cdot + \cdot H \longrightarrow H_2$$

The covalent bond in H_2 is represented by a single line joining the two H atoms: H—H. (Covalent bonds are sometimes shown as two dots instead of a line, like H:H, but we will use lines for covalent bonds to emphasize the difference between the electrons in a bond and pairs of electrons that are not involved in bonding.) Thus, the lone electrons of the two hydrogen atoms are paired to form a covalent bond. Although the electrons would repel each other because they bear the same charge, they are also attracted by the positively charged nuclei. This attractive force is the "chemical glue" that holds atoms together and is the basis of stability in molecules.

Inspection of the dot symbols for O, N, and C in Figure 7.4 reveals that for each of them to feel like it has as many electrons as neon (the noble gas that follows them

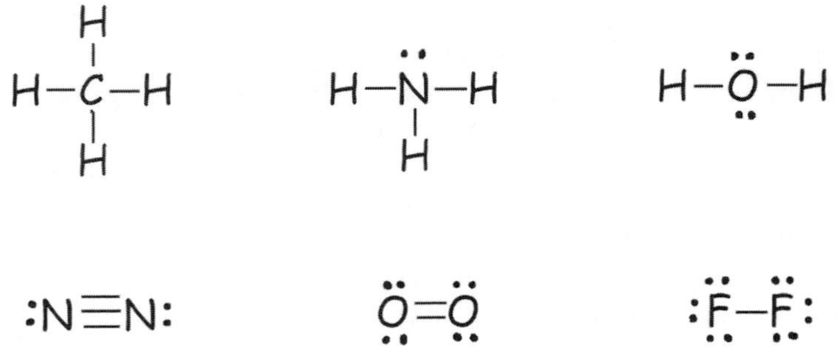

Figure 7.5   Lewis structures of some simple molecules

$$
\begin{array}{ccc}
& \text{H} & \\
& | & \\
\text{H}-\text{C}-\text{H} & \quad \text{H}-\ddot{\text{N}}-\text{H} \quad & \text{H}-\ddot{\text{O}}-\text{H} \\
& | & \\
& \text{H} & \quad | \\
& & \quad \text{H}
\end{array}
$$

$$:\text{N}{\equiv}\text{N}: \qquad \ddot{\text{O}}{=}\ddot{\text{O}} \qquad :\ddot{\text{F}}{-}\ddot{\text{F}}:$$

at the end of the second period), oxygen would have to form two bonds (as it does in water), nitrogen would have to form three bonds (as it does in ammonia), and carbon would have to form four bonds (as it does in methane) (Figure 7.5). The dot symbols also show the valence electrons that are not involved in bonding.

The structures shown in Figure 7.5 are called Lewis structures. A **Lewis structure** is a representation of covalent bonding in which shared electron pairs are shown as lines and electrons that are not shared (called nonbonding electrons or **lone pairs**) are shown as pairs of dots on individual atoms. For example, in the Lewis structure of water, the O atom has two lone pairs, while the H atom has no lone pairs because its only electron is used to form a covalent bond.

There are some simple rules for writing Lewis structures that we will describe later in the chapter. Although these rules enable you to predict the bonding arrangement in molecules, structures with lines and dots don't provide a complete description of bond formation. For example, consider the Lewis structures for O_2 and N_2 in Figure 7.5, both of which are constructed by simply pairing the unpaired dots. To pair all the unpaired dots, you need to make two bonds in O_2 and three bonds in N_2. What does it mean to have three bonds? Is it possible to have four bonds between two atoms? How many bonds can one atom have? In order to understand the answers to these questions, you have to learn something more about covalent bonds. Understanding the nature of covalent bonds—how they're formed and what they look like—will help you visualize molecules so that when you draw Lewis structures, you can think of them as representing real objects, not just a bunch of lines and dots.

What Is a Covalent Bond?

Representing a covalent bond with a single line between two atoms indicates, in the simplest possible way, a connection between the two atoms, but it doesn't tell us anything about the nature of the connection. What exactly is a covalent bond? We know it contains two electrons, and these electrons are being shared by two atoms, each with its own positively charged nucleus. It should be fairly intuitive that the electrons will occupy a space between the two nuclei. In Chapter 6 you learned about spaces occupied by electrons—they're called *atomic orbitals,* and you have seen that quantum mechanics describes their shapes. If two electrons are in a region of space within a molecule, this is also called an orbital, but it's called a *molecular orbital.* So if a covalent bond is two electrons shared by two atoms in a space between the nuclei and if a molecular orbital is the region of space occupied by two electrons shared between two nuclei, it follows that a covalent bond can be described as a molecular orbital containing two electrons. The good news is that there are only two types of molecular orbitals, which we will now describe in more detail.

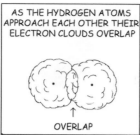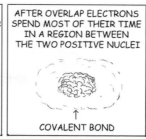

Figure 7.6 Two hydrogen atoms overlap to form a covalent bond.

Covalent Bonds Form When Atomic Orbitals Overlap

To better understand the nature of a covalent bond, let's imagine how it forms between two hydrogen atoms. Imagine two hydrogen atoms approaching each other as shown in Figure 7.6. Both atoms have their lone electron in a 1s atomic orbital. Now imagine the orbitals as spherical clouds, with a positive proton in the center (remember the nucleus of hydrogen is just a proton). Because they are clouds (like rain clouds or dust clouds), the orbitals would not stop when they touch each other; instead they would overlap and blend together such that there would be a region of space shared by both clouds. Once the hydrogen clouds have overlapped, you can imagine that an electron in this overlapped region would feel the force of attraction from both nuclei. This is the essence of the formation of a covalent bond, that is, such a bond forms when the atomic orbitals of valence electrons overlap. The idea is that there is an optimal overlap between the two atomic orbitals where there is a balance between the attraction of the electrons and the two nuclei and the repulsion between the two nuclei and two electrons. This overlap gives rise to the covalent bond—the electrons are now most likely to be found in the region between the two nuclei.

Electrons in Covalent Bonds Occupy Molecular Orbitals

The theory that describes the space occupied by electrons in a covalent bond in terms of molecular orbitals is called **molecular orbital theory**. As shown in Figure 7.6, the electron density, which corresponds to the molecular orbital formed in H_2, looks like a sausage between the two nuclei. This is called a *sigma molecular orbital* and placing two electrons in such an orbital gives rise to a *sigma bond*. To help you remember this, sigma (σ) is the Greek letter for s, the first letter in sausage. The situation is simplest for the hydrogen molecule. Two 1s orbitals overlap and form a sigma

**NO ONE REALLY UNDERSTANDS THIS SEEMINGLY MAGICAL
TRANSFORMATION OF ATOMIC ORBITALS TO MOLECULAR ORBITALS**

Two overlapped s orbitals become a sigma bond.

Figure 7.7 The shapes of *p* orbitals

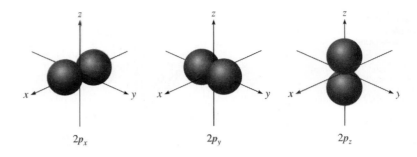

$2p_x$ $2p_y$ $2p_z$

molecular orbital. It gets a little more complicated when we describe bonds that are formed from the overlap of *p* orbitals and when we describe molecules that have double and triple bonds. But remember this about covalent bond formation: **Atomic orbitals each containing an electron overlap and change into a molecular orbital containing a pair of electrons.** The transformation of atomic orbitals to molecular orbitals is a quantum mechanical process, the details of which need not concern us here.

Most bonds are formed from the overlap of *p* orbitals, which you know from Chapter 6 to have the dumbbell shapes shown in Figure 7.7. Let's consider molecular fluorine as another example. The electron configuration of F is $1s^2 2s^2 2p^5$. The valence electrons in this case are in the second shell (principal quantum number $n = 2$), so there are a total of seven valence electrons on each F atom (two 2*s* and five 2*p* electrons). Because one of the *p* orbitals contains an unpaired electron, two F atoms will readily react to form an F_2 molecule, in which one covalent bond—a sigma bond, that is, a sigma molecular orbital containing the two electrons from the 2*p* orbitals—is formed.

$$:\ddot{F}\cdot + \cdot\ddot{F}: \longrightarrow :\ddot{F}:\ddot{F}: \quad \text{or} \quad :\ddot{F}-\ddot{F}:$$

To understand how two *p* orbitals can form a sigma bond, let's consider two fluorine atoms, each with an electron in its $2p_x$ orbital, approaching the other F atom along the *x* axis. Because the *p* orbital lies along the internuclear axis (as we have arbitrarily defined it), the orbitals will overlap as shown in Figure 7.8. Now, just as we did for the overlap of *s* orbitals, imagine the *p* orbitals changing into a sigma molecular orbital. Thus, two p_x orbitals can overlap to form a sigma bond. What happens to the rest of the valence electrons already paired up in atomic orbitals? These lone pairs of electrons stay on the individual F atoms in 2*s*, $2p_y$, and $2p_z$ orbitals, which also must undergo some sort of shape change when the molecule forms. We will simply refer to these as nonbonding or lone pair orbitals.

What other types of orbitals can form sigma bonds? Consider the formation of the HF molecule. As the H and F atoms approach each other, imagine the 1*s* orbital of H overlapping with the $2p_x$ orbital of F, and (poof) a sigma molecular orbital is formed. Again, we have defined the *x* axis as the internuclear axis.

Figure 7.8 Two fluorine atoms overlap to form a covalent bond.

THE LONE ELECTRON IN THE 2 Px ORBITAL OF FLUORINE FORMS A DUMBBELL-SHAPED CLOUD AROUND THE NUCLEUS

F   F

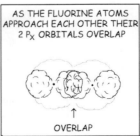

AS THE FLUORINE ATOMS APPROACH EACH OTHER THEIR 2 Px ORBITALS OVERLAP

OVERLAP

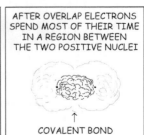

AFTER OVERLAP ELECTRONS SPEND MOST OF THEIR TIME IN A REGION BETWEEN THE TWO POSITIVE NUCLEI

COVALENT BOND

Two overlapped p_x orbitals become a sigma bond.

Overlapped s and p_x orbitals become a sigma bond.

So we've seen that two s orbitals can overlap to form a sigma bond, two p_x orbitals can overlap to form a sigma bond, and an s and a p_x can also form a sigma bond. Note also that these orbitals can come from any shell. For example, $2p_x$ and $4p_x$ orbitals can overlap, $1s$ and $2p_x$ orbitals can overlap, $3p_x$ and $4s$ orbitals can overlap, and so on. Any of these combinations will produce a sigma bond, but sigma bonds cannot be formed from the overlap of p_y and p_z orbitals because of their different orientations.

To understand how electrons in p_y and p_z orbitals can be involved in bonding, we have to consider the second type of molecular orbital, the one that is involved in double and triple bonds. Let's consider the formation of molecular oxygen, O_2. The electron configuration of O is $1s^2 2s^2 2p^4$, so there are a total of six valence electrons on each O atom (two $2s$ and four $2p$ electrons). Because two of the p orbitals contain unpaired electrons, we expect the two O atoms will react to form an O_2 molecule with two covalent bonds.

$$\cdot \ddot{O} \cdot + \cdot \ddot{O} \cdot \longrightarrow \ddot{O} = \ddot{O}$$

How does each O atom form two bonds? Consider two O atoms, each with an electron in the $2p_x$ and $2p_y$ orbital, approaching each other along the x axis. While we expect the $2p_x$ orbitals to overlap to form a sigma bond, it's not obvious what happens to the $2p_y$ orbitals. Imagine the orbitals continuing their movement toward each other until the $2p_y$ orbitals overlap as shown in Figure 7.9. The orbitals make contact at two places above and below the sigma bond. At this point, the two $2p_y$ orbitals change into a molecular orbital in which the electron density is concentrated in two lobes above and below the sigma molecular orbital. (They're pretty weird, but a lot of things about electrons are weird.)

Figure 7.9 The two 2*p* orbitals on oxygen atoms overlap to form a double bond.

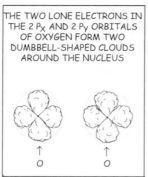

THE TWO LONE ELECTRONS IN THE 2 Pₓ AND 2 Pᵧ ORBITALS OF OXYGEN FORM TWO DUMBBELL-SHAPED CLOUDS AROUND THE NUCLEUS

↑ O   ↑ O

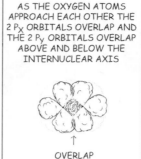

AS THE OXYGEN ATOMS APPROACH EACH OTHER THE 2 Pₓ ORBITALS OVERLAP AND THE 2 Pᵧ ORBITALS OVERLAP ABOVE AND BELOW THE INTERNUCLEAR AXIS

↑ OVERLAP

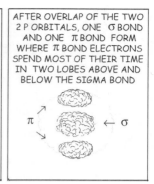

AFTER OVERLAP OF THE TWO 2 P ORBITALS, ONE σ BOND AND ONE π BOND FORM WHERE π BOND ELECTRONS SPEND MOST OF THEIR TIME IN TWO LOBES ABOVE AND BELOW THE SIGMA BOND

π   ← σ

Two overlapped p_y orbitals become a pi bond.

The molecular orbital formed by the sideways overlap of *p* orbitals is called a pi molecular orbital. Pi bonds usually accompany sigma bonds, and they normally occur in molecules with multiple (double or triple) bonds. Anytime there is a single bond between two atoms, it will most likely be a sigma bond, not a pi bond. To help you remember pi bonds, imagine the pi symbol, π, where the two legs represent the two lobes of the orbital and the bar across the top signifies the sigma bond that must also be present.

At this point, you may already suspect the origin and nature of the triple bond. Consider the formation of molecular nitrogen, N_2. The electron configuration of N is $1s^2 2s^2 2p^3$, so there are a total of five valence electrons on each N atom (two 2*s* and three 2*p* electrons). Because all three of the 2*p* orbitals contain unpaired electrons, two N atoms will readily react to form an N_2 molecule with three covalent bonds.

$$\cdot \ddot{N} \cdot \; + \; \cdot \ddot{N} \cdot \; \longrightarrow \; :N \equiv N:$$

In this case, the two N atoms have unpaired electrons in their $2p_x$, $2p_y$, and $2p_z$ orbitals. Approaching each other along the *x* axis, the $2p_x$ orbitals overlap to form a sigma bond, the $2p_y$ orbitals overlap to form a pi bond, and the $2p_z$ orbitals will overlap in the same way that the $2p_y$ orbitals do except they will be 90° away from the $2p_y$ orbitals. Thus, the $2p_z$ orbitals overlap to form a second pi bond, as shown in Figure 7.10. Whenever a molecule contains a triple bond, it comprises one sigma bond and two pi bonds, or a total of six electrons in three molecular orbitals.

As you might expect, triple bonds are stronger than double bonds, and double bonds are stronger than single bonds. That is, it takes more energy to pull the atoms apart if there are more covalent bonds—there is simply more chemical glue holding them together. Another feature of multiple bonds is that they are shorter than single bonds when atoms of similar sizes are involved. The additional chemical glue pulls

THE THREE LONE ELECTRONS IN THE 2 P_X, 2 P_y, AND 2 P_z ORBITALS OF NITROGEN FORM THREE DUMBBELL-SHAPED CLOUDS AROUND THE NUCLEUS	AS THE NITROGEN ATOMS APPROACH EACH OTHER THE 2 P_X ORBITALS OVERLAP, THE 2 P_y ORBITALS OVERLAP, AND THE 2 P_z ORBITALS OVERLAP	AFTER OVERLAP, ONE σ BOND AND TWO π BONDS FORM WHERE EACH π BOND HAS TWO LOBES FLANKING THE σ BOND IN THE CENTER

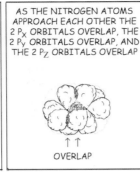

N   N

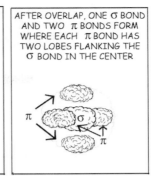

↑↑

OVERLAP

Figure 7.10 The three $2p$ orbitals on nitrogen atoms overlap to form a triple bond.

the atoms closer together. Table 7.1 shows the lengths of some common bonds and the amount of energy needed to break these bonds on a molar basis.

So far, the description of chemical bond formation in terms of orbital overlaps works well for diatomic molecules. The situation is not quite so straightforward for polyatomic molecules. Consider the methane (CH_4) molecule. How are the four C—H bonds formed? The valence electron configuration of carbon is $2s^2 2p^2$, so there are only two unpaired electrons in the $2p$ orbitals (say, $2p_x$ and $2p_y$). We would therefore expect carbon to form only two sigma bonds with hydrogen, resulting in a CH_2 molecule. In order for carbon to form four sigma bonds with hydrogen, the two $2s$ electrons must first become unpaired, and then the four unpaired electrons are distributed among the $2s$ and three $2p$ orbitals. The source of energy needed for this to happen is part of the activation energy, that is, energy input to promote a chemical reaction. To account for the fact that the four C—H bonds are equivalent in length and strength, imagine an additional step in which the four orbitals are mixed or "hybridized" to generate four equivalent orbitals. These orbitals are called sp^3 (pronounced s-p three) hybrid orbitals because they arise as a result of the mixing of one s and three p orbitals. The only difference between these orbitals is that they point in different directions. (*See website for a general discussion of hybridization in chemistry.*)

Table 7.1 Average Bond Lengths and Bond Energies for Some Common Single, Double, and Triple Bonds		
Bond Type	**Bond Length (pm)***	**Bond Energy (kJ/mol)**
C—H	107	414
C—O	143	351
C=O	121	799
C—C	154	347
C=C	133	620
C≡C	120	812
C—N	143	276
C=N	138	615
C≡N	116	891
N—O	136	176
N=O	122	458
O—H	96	460

*1 pm = 1×10^{-12} m

The carbon atom can readily form single, double, and triple bonds. In methane (CH_4), the C atom forms four single bonds with H atoms. In ethylene (C_2H_4), each C atom forms a double bond with another carbon atom and two single bonds with hydrogen. Carbon can also form a triple bond, as in acetylene (C_2H_2), where each carbon atom forms a triple bond with the other carbon atom and a single bond with a hydrogen atom. What kinds of molecular orbitals are involved in these bonds? Figure 7.11 shows Lewis structures for ethylene and acetylene and pictures of the molecular orbitals.

Figure 7.11 Lewis structures and orbital pictures for ethylene and acetylene

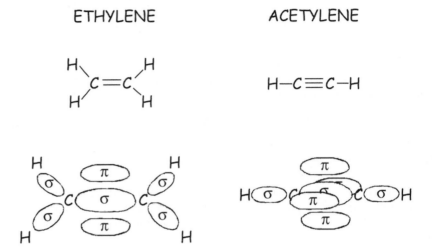

Lewis Structures | *Depicting molecules with lines and dots*

Now that you understand the nature of covalent bonds, we will return to our discussion of Lewis structures. Based on his idea that atoms form covalent bonds to achieve a noble gas electron configuration, Lewis formulated the **octet rule** to guide the drawing of Lewis structures. According to this rule, an atom other than hydrogen tends to form bonds until it is surrounded by eight valence electrons. We imagine that each of the two atoms joined in a covalent bond feels like it has both of the electrons. By sharing electrons in covalent bonds, the individual atoms can complete their octets. Reviewing the Lewis structures of CH_4, NH_3, H_2O, N_2, O_2, and F_2, we see that the C, N, O, and F atoms all satisfy the octet rule. In all of the preceding cases involving multiple bonds, the octet rule is satisfied for the C, N, and O atoms. Note that Lewis structures also apply to polyatomic ions in which the atoms are held together by covalent bonds.

We hope that you will eventually be able to draw Lewis structures based on your knowledge of the way specific atoms form bonds. In the meantime, the following steps and hints will help you draw Lewis structures for any molecule or polyatomic ion in a systematic way.

1. **Write the skeletal structure of the compound.** For simple compounds where there is a central atom surrounded by one or more atoms, this is pretty straightforward. Arrange the chemical symbols for the atoms such that a central atom is surrounded by the other atoms and draw a single line between the central atom and each of the surrounding atoms. In general, the least electronegative atom occupies the central position, and hydrogen and fluorine always occupy the terminal (end) positions. For more complex compounds, you either need more information than just the chemical formula or you make an intelligent guess. To make an intelligent guess, you'll need to become familiar with patterns of

bonding for specific atoms. The good news is that in the vast majority of cases, you have to do this for only H, C, N, O, F, P, S, and Cl. It's easy to predict how the rest of the nonmetallic elements will bond because atoms in the same group often bond in a similar way. For example, Br and I form bonds with other atoms exactly like chlorine does. We will come back to bonding patterns later.

2. **Count the total number of valence electrons present,** keeping in mind that this number is equal to the element's group number. For polyatomic anions, add the number of negative charges to the total number. (For example, for the CO_3^{2-} ion, we add two electrons to the total number of valence electrons in one C and three O atoms because the -2 charge indicates there are two more electrons than are provided by the atoms.) For polyatomic cations, we subtract the number of positive charges from this total. (Thus, for the NH_4^+ ion, we subtract one electron from the total number of valence electrons because the $+1$ charge indicates a loss of one electron from the group of neutral atoms.)

3. **Subtract two electrons for each bond from the total number of valence electrons and distribute the remaining electrons as lone pairs around the atoms to achieve an octet for each atom (except H).**

4. **If the octet rule is not satisfied for all atoms, try adding double or triple bonds between the surrounding atoms and the central atom, using the lone pairs to form additional bonds.**

Of course, the only way you will learn to draw Lewis structures is practice, practice, practice. Let's try some examples.

7.1 EXAMPLE

Draw the Lewis structure for the NF$_3$ (nitrogen trifluoride) molecule.

Answer
Step 1: *Draw the skeletal structure and place N in the center because it's less electronegative than F. Remember that the F atom always occupies an end position.*

$$F-N-F$$
$$|$$
$$F$$

Step 2: *The outer-shell electron configurations of N and F are $2s^2 2p^3$ and $2s^2 2p^5$, respectively. Therefore, N has 5 valence electrons and F has 7 valence electrons, and there are a total of $3 \times 7 + 5$ or 26 electrons to account for. Of course, you could also get the number of valence electrons directly from the group numbers on the periodic table.*
Step 3: *Subtract 6 electrons for the three N—F bonds from the 26 and distribute the remaining electrons, trying to complete the octets for the F and N atoms. It's usually best to try filling the octets of the surrounding atoms first before the central atom.*

$$:\ddot{F}-\ddot{N}-\ddot{F}:$$
$$|$$
$$:\ddot{F}:$$

Because this structure satisfies the octet rule for all the atoms, Step 4 is not required. To check, count the valence electrons in NF$_3$ (in covalent bonds and in lone pairs). The result is 26, the same as the total number of valence electrons we had to start with.

EXAMPLE

7.2

Draw the Lewis structure for HNO$_3$ (nitric acid).

Answer
Step 1: In drawing the skeletal structure, we note that N is less electronegative than O, so it should occupy the central position, and that H occupies an end position. If there is a choice, H will often be bonded to the more electronegative atom, which is O in this case:

$$H-O-N-O$$
$$\,|$$
$$O$$

Step 2: The number of valence electrons in N, O, and H are 5, 6, and 1, respectively, corresponding to their periodic group numbers. Therefore, there are a total of 5 + (3 × 6) + 1 or 24 valence electrons to account for.
Step 3: Subtract 8 electrons for the four bonds and distribute the remaining electrons to try to complete the octets for O first, then N:

$$H-\ddot{O}-N-\ddot{O}:$$
$$|$$
$$:\ddot{O}:$$

There are not enough electrons to give every atom (but H) an octet—nitrogen only has six electrons around it—so we need to go to Step 4.
Step 4: Move a lone pair from one of the end O atoms to form another bond with N. Now the octet rule is also satisfied for the N atom:

$$H-\ddot{O}-N=\ddot{O}$$
$$|$$
$$:\ddot{O}:$$

Thus, the Lewis structure that satisfies the octet rule for N and O predicts a double bond between N and O. What does this double bond look like? If we draw the molecular orbitals, we can depict what it would look like, as in Figure 7.12.

It turns out that there is a problem with this picture. According to Table 7.1, we would expect that the N=O bond would be shorter and stronger than the N—O bond. However, that's not the case. Experiments show that both N—O bonds (the two that don't have H bonded to the O) are equal in length and strength. In fact, they have lengths and strengths that are intermediate between a single and a double bond, while that of the N—O bond in N—O—H is equal to a single bond. How can we explain this unusual bonding behavior?

Figure 7.12 Picture of molecular orbitals in nitric acid if there is one double bond

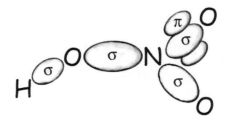

Delocalized Pi Orbitals

Pi bonds can spread out

The explanation for the bonding in HNO_3 is that the pi orbitals spread out over the N and two O atoms. You can understand how this might happen if you imagine the $2p_y$ orbitals of the N and O atoms, each with one electron in it, overlapping to form two banana-shaped regions covering all three atoms above and below the sigma bond (Figure 7.13).

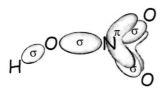

Figure 7.13 Picture of delocalized pi molecular orbitals in nitric acid

According to our rules for orbitals, we are likely to find the three valence electrons somewhere in this spread-out region, which we refer to as *delocalized orbitals*. Recall that an orbital is defined as a space where there is a 90% chance of finding an electron at any time. In this unusual case, there are three electrons in a space double the size of a normal pi orbital. We say that the electrons are delocalized because they are not confined between two atoms like in normal bonds. Delocalized orbitals also make molecules more stable because the chemical glue is distributed over more atoms. How do we know if a molecule has delocalized electrons? It turns out that it is often apparent when we draw the Lewis structure. But how can we represent delocalized electrons with a Lewis structure?

Resonance

Moving electrons in Lewis structures

Because we are limited to lines between two atomic symbols for covalent bonds, a single Lewis structure cannot represent the actual bonding arrangement in molecules with delocalized pi orbitals. Faced wih this dilemma, chemists devised a procedure called **resonance,** which means using two or more Lewis structures to represent a particular molecule or ion. Figure 7.14 shows two **resonance structures** for nitric acid. Note that the two structures have the same arrangement of atoms. **Atoms must not be moved when drawing resonance structures; only electron pairs can be moved.** The structures in Figure 7.14 are called *equivalent resonance structures* because they have the same number of bonds to the same atoms. The idea is that the actual structure is an average or composite of the two resonance structures. For example, if you averaged the two N—O bonds without H attached from the two structures, they would each have the equivalent of 1.5 bonds. This is essentially the case when the pi bond

Figure 7.14 Resonance structures for nitric acid

is delocalized around both N—O sigma bonds. The double-headed arrow is used to indicate that the structures shown are resonance structures.

A common misconception about resonance is that a molecule or ion somehow shifts quickly back and forth from one resonance structure to the other. This does not happen. Each resonance structure is a nonexistent species. The following analogy may help you understand the concept of resonance. A medieval European traveler to Africa returns home and describes a rhinoceros as a cross between a griffin and a unicorn, two familiar-looking but imaginary animals. In the same way, we describe the nitric acid molecule, a real species, in terms of two familiar-looking but nonexistent structures. Of course, the picture with the banana-shaped pi orbital is the rhinoceros in this case, but we can't depict it with a single Lewis structure.

We said earlier that it is often apparent from the Lewis structure if a molecule has delocalized electrons. Let's look at another example to see how this works.

7.3

EXAMPLE

Draw the Lewis structure for CO_3^{2-} (carbonate ion).

Although the carbonate ion is a charged species, the atoms within the ion are joined by covalent bonds. For this reason, we can draw a Lewis structure for the ion just as we do for molecules.

Answer

Step 1: *You can deduce the skeletal structure of the carbonate ion by recognizing that C is less electronegative than O. Therefore, it is most likely to occupy a central position as follows:*

$$O - C - O \\ \overset{|}{O}$$

Step 2: *The number of valence electrons in C and O are 4 and 6, respectively, corresponding to their group numbers, and the ion has two negative charges. Therefore, the total number of electrons is 4 + (3 × 6) + 2 or 24.*

Step 3: *Subtract 6 electrons for the three bonds and distribute the remaining electrons to try to complete octets for O first, then C:*

$$\left[:\ddot{O} - C - \ddot{O}: \atop :\ddot{O}: \right]^{2-}$$

This structure shows all 24 electrons [9 lone pairs (18 electrons) and 3 bonds (6 electrons)]; however, the octet rule is not satisfied for C. We must go to Step 4.

Step 4: *Move a lone pair from one of the O atoms to form another bond with C. Now the octet rule is also satisfied for the C atom:*

$$\left[:\ddot{O} - C - \ddot{O}: \atop :O: \right]^{2-}$$

By now, it should have occurred to you that the double bond could be drawn to any of the oxygens, and all of the structures would be equivalent. Whenever

—Continued next page

Continued—

this is the case, you should draw resonance structures. For the carbonate ion, you can draw three of them:

$$\left[\begin{array}{c} :\!\overset{..}{O}\!: \\ \| \\ :\!\overset{..}{O}\!-\!C\!-\!\overset{..}{O}\!: \\ \overset{..}{} \end{array} \right]^{2-} \longleftrightarrow \left[\begin{array}{c} :\!\overset{..}{O}\!: \\ | \\ \overset{..}{O}\!=\!C\!-\!\overset{..}{O}\!: \\ \overset{..}{} \end{array} \right]^{2-} \longleftrightarrow \left[\begin{array}{c} :\!\overset{..}{O}\!: \\ | \\ :\!\overset{..}{O}\!-\!C\!=\!\overset{..}{O} \\ \overset{..}{} \end{array} \right]^{2-}$$

What do you think the carbonate ion would look like? The Lewis structure indicates that the pi orbital is delocalized over the entire ion. This turns out to be the case, and Figure 7.15 shows the $2p_y$ orbitals involved and the delocalized orbital formed by them.

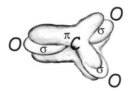

Figure 7.15 Picture of delocalized pi molecular orbitals in carbonate ion

The useful feature about resonance structures is that they suggest the possibility of delocalized pi orbitals in molecules. If you can draw equivalent Lewis structures by moving double bonds around, it's a good indication that the pi electrons are delocalized in the actual molecule or ion.

At this point, there are several questions we could ask: Are all resonance structures equivalent? If not, are some resonance structures better than others? If so, how can we tell? To answer these questions, we have to introduce you to one more feature of Lewis structures—the concept of formal charge.

Formal Charge and Lewis Structures

Did you notice anything unusual about the Lewis structure for nitric acid? There are four bonds to nitrogen; however, we told you that N forms three bonds and has a lone pair, which is consistent with its having five valence electrons. What are the

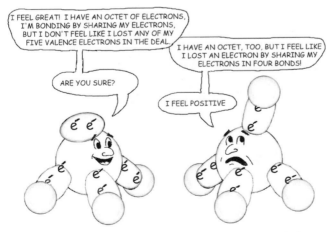

Nitrogen with three bonds and a lone pair has no formal charge, but nitrogen with four bonds and no lone pair has a formal charge of $+1$.

implications of a nitrogen atom with four bonds? Compared to the way N normally bonds, with three of its valence electrons shared in three bonds and one lone pair of electrons all to itself, the N with four bonds and no lone pairs seems to be short of one electron. If it actually were one electron short, it would have a +1 charge.

The concept of **formal charge,** which is simply assigning charges according to how valence electrons are used (like we did earlier), is a way to keep track of how the valence electrons are distributed in Lewis structures. An atom's formal charge is the electrical charge difference between the number of valence electrons of an atom and the number of electrons assigned to that atom in a Lewis structure. For purposes of bookkeeping, we assign all of the bonding electrons and lone pairs in a molecule to specific atoms. This means we have to imagine splitting the bonds and giving one electron to each of the bonded atoms. Any lone pairs of electrons on atoms are assigned to them solely as their own.

Thus, to assign the number of electrons on a particular atom in a Lewis structure, we simply add all the nonbonding (or lone pair) electrons and one electron for each covalent bond formed by the atom. For the nitric acid molecule, the formal charge on nitrogen is +1 because it started with 5 valence electrons and is assigned only 4 electrons from the Lewis structure, one electron for each bond. The formal charge is the number of valence electrons minus the number of assigned electrons, which in this case is 5 − 4 or +1. For the oxygen with one bond to nitrogen, the formal charge is 6 − 7 or −1. For single positive and negative charges, we normally omit the numeral 1 as shown in Figure 7.16. The formal charges on the other two O atoms are 6 − 6 or zero.

Note that the sum of the formal charges is equal to zero for nitric acid because it's a neutral molecule. For cations, the sum of the formal charges must equal the positive charge on the cation. For anions, the sum of the formal charges must equal the

Figure 7.16 Resonance structures for nitric acid with formal charges

negative charge on the anion. Remember that nitrogen with four bonds and no lone pairs will always have a $+1$ formal charge and oxygen with one bond and three lone pairs will always have a -1 formal charge. These are good rules to know because they occur fairly often.

Let's review the process of determining formal charge using the ozone molecule (O_3) as an example. Proceeding by steps, as we did earlier, we arrive at the following two equivalent Lewis resonance structures:

$$\ddot{O}=\ddot{O}-\ddot{O}: \longleftrightarrow :\ddot{O}-\ddot{O}=\ddot{O}$$

The formal charge on each atom in O_3 can now be calculated according to the following scheme,

	$\ddot{O}$	$\ddot{O}$	$\ddot{O}:$
Valence e^-	6	6	6
e^- assigned to atom	6	5	7
Difference (formal charge)	0	$+1$	-1

where the wavy lines denote the breaking of the bonds. Note that the breaking of a double bond results in a transfer of two electrons to each of the bonding atoms, one for each covalent bond. Thus the formal charges of the atoms in O_3 are

$$\overset{+}{\ddot{O}}=\overset{-}{\ddot{O}}-\ddot{O}: \longleftrightarrow :\overset{-}{\ddot{O}}-\overset{+}{\ddot{O}}=\ddot{O}$$

Keep in mind that formal charges don't represent actual charge separation within the molecule. In the O_3 molecule, for example, there is no evidence that the central atom bears a net $+1$ charge or that one of the end atoms bears a -1 charge. Writing these charges on the atoms in the Lewis structure merely helps us keep track of the valence electrons in the molecule. In this case, the pi electrons would be delocalized so the electron density would be the same at the two end O atoms.

Now let's get back to the questions we raised earlier.

1. Are all resonance structures equivalent? The answer is no (although so far, all of the resonance structures we have drawn are equivalent).
2. If not, are some resonance structures better (that is, chemically more plausible) than others? The answer is yes.
3. If so, how can we tell? The better resonance structures are usually the ones with minimal formal charge. Resonance structures having large formal charges (equal to or greater than -2 or $+2$) are less plausible, meaning that those structures contribute less to the overall properties of the molecule.

Draw three resonance structures of the nitrous oxide molecule (N_2O) in which the atoms are arranged as NNO. Rank the resonance structures in increasing order of importance.

7.4

EXAMPLE

Answer   *By following the preceding procedures, we arrive at a Lewis structure for N_2O*

$$(A) \qquad \overset{-}{\ddot{N}}=\overset{+}{N}=\ddot{O}$$

However, we can draw another Lewis strcture like

Continued—

(B) $:N\equiv\overset{+}{N}-\overset{..}{\underset{..}{O}}:^{-}$

Note that all we have done is to shift the electrons so that there is now a triple bond between the N atoms and a single bond between N and O. We can draw yet a third resonance structure:

(C) $^{2-}\overset{..}{\underset{..}{:N}}-\overset{+}{N}\equiv\overset{+}{O}:$

How do we rank these three resonance structures? (C) should be the least plausible because it involves a large formal charge (−2 on the end N). There is not much to choose between (A) and (B) except that the negative charge is on the more electronegative O atom in (B), as it should be. Therefore, we rank these resonance structures in increasing order of importance as

(C) < (A) < (B)

(Check the website for more problems on writing resonance structures.)

Exceptions to the Octet Rule

The octet rule works particularly well for elements in the second period of the periodic table. This is because the atoms of these elements have $2s$ and $2p$ subshells, accounting for a maximum of eight electrons. When an atom has fewer than eight electrons, it would pair up with some other atom or atoms to achieve the octet configuration. But there are three exceptions to the octet rule that you should know. In particular, you should also know that the first two apply only to a limited number of molecules, while the third, the expanded octet, is quite common in many molecules.

1. The Incomplete Octet

In some compounds, the number of electrons surrounding the central atom in a stable molecule is fewer than eight. Consider, for example, beryllium, which is a Group 2A (and a second-period) element with two valence electrons in the $2s$ orbital. In the gas phase, beryllium hydride (BeH_2) exists as discrete molecules. The Lewis structure of BeH_2 is

H—Be—H

As you can see, only four electrons surround the Be atom, and there is no way to satisfy the octet rule for beryllium in this molecule.

Elements in Group 3A, particularly boron and aluminum, also tend to form compounds in which they are surrounded by fewer than eight electrons. Boron, with three valence electrons $(2s^2 2p^1)$, reacts with the halogens to form a class of compounds having the general formula BX_3, where X is a halogen atom. Thus, in boron trifluoride, there are only six electrons around the boron atom:

$:\overset{..}{\underset{..}{F}}-B-\overset{..}{\underset{..}{F}}:$
$\quad\quad\overset{|}{\underset{..}{:F:}}$

The only elements that may form compounds with incomplete octets are Be, B, and Al. (Note that it is possible for metals such as Be and Al to form covalent compounds.)

2. Odd-Electron Molecules

Some molecules contain an *odd* number of electrons. Among them are nitric oxide (NO) and nitrogen dioxide (NO_2):

$$\ddot{N}\!=\!\ddot{O} \qquad \ddot{O}\!=\!\overset{+}{N}\!-\!\ddot{O}\!:^{-}$$

Because we need an even number of electrons for complete pairing (to reach eight), the octet rule clearly cannot be satisfied for all the atoms in any of these molecules. There are very few stable molecules with an odd number of electrons.

3. The Expanded Octet

As mentioned earlier, atoms of the second-period elements cannot have more than eight valence electrons around the central atom, but atoms of elements in and beyond the third period of the periodic table readily form compounds in which more than eight electrons surround the central atom. How can this happen? We have seen that the first-period element hydrogen can have only two electrons because the first shell is $1s$. The second shell has two subshells of $2s$ and $2p$, which together limit the number of electrons around second-period elements to eight. How many electrons can the third shell hold? In addition to the $3s$ and $3p$ subshells, these elements in the third period also have the $3d$ subshell that can be used in bonding. The presence of a $3d$ subshell enables an atom to form an **expanded octet.** The commonly encountered elements that form expanded octets are P, S, Cl, Br, and I.

One compound in which there is an expanded octet is sulfur hexafluoride (SF_6), a very stable compound. In SF_6, each of sulfur's six valence electrons forms a covalent bond with a fluorine atom, so there are 12 electrons around the central sulfur atom:

$$
\begin{array}{ccc}
 & :\!\ddot{F}\!: & \\
:\!\ddot{F} & \!\!\diagdown\!\!\mid\!\!\diagup\!\! & \ddot{F}\!: \\
 & S & \\
:\!\ddot{F} & \!\!\diagup\!\!\mid\!\!\diagdown\!\! & \ddot{F}\!: \\
 & :\!\ddot{F}\!: & \\
\end{array}
$$

Keep in mind that sulfur also forms many compounds in which it obeys the octet rule. In these cases, it often bonds like oxygen to form two bonds and two lone pairs as in sulfur dichloride:

$$:\!\ddot{C}l\!-\!\ddot{S}\!-\!\ddot{C}l\!:$$

EXAMPLE 7.5

Draw the Lewis structure of phosphorus pentachloride (PCl_5).

Answer *The outer-shell electron configurations of P and Cl are $3s^2 3p^3$ and $3s^2 3p^5$, respectively, and so the total number of valence electrons is $5 + (5 \times 7) = 40$. Phosphorus is a third-period element, and therefore it can have an expanded octet. The Lewis structure of PCl_5 is*

$$
\begin{array}{ccc}
 & :\!\ddot{C}l\!: & \\
:\!\ddot{C}l & \!\!\diagdown\!\!\mid\!\! & \\
 & P\!-\!\ddot{C}l\!: & \\
:\!\ddot{C}l & \!\!\diagup\!\!\mid\!\! & \\
 & :\!\ddot{C}l\!: & \\
\end{array}
$$

Note that there are five bonds and hence 10 valence electrons around the P atom.

Table 7.2 Bonding Patterns for C, N, and O

Bonding Pattern	Examples

Now you're ready to try your hand at drawing Lewis structures on your own. Remember, the only way you will learn to draw Lewis structures is practice, and there are lots of examples on the website. As you practice drawing Lewis structures, try to look for patterns of bonding. You already know that H, O, N, and C only form 1, 2, 3, and 4 bonds, respectively (that is, when they don't have a formal charge). Table 7.2 shows examples of Lewis structures for molecules containing C, N, and O that illustrate their typical bonding patterns. Note that O has two bonds and two lone pairs and N has three bonds and one lone pair. This will always be the case unless the atom has a formal charge. You have seen examples of this in CO_3^{2-} and HNO_3, where O with one bond and three lone pairs has a formal charge of -1 and N with four bonds and no lone pairs has a formal charge of $+1$.

In addition to the normal ways in which specific atoms bond, there are patterns that apply to certain types of molecules such as oxoacids. As we saw in Chapter 2, there are many oxoacids (like nitric acid) and oxoanions (like carbonate). It turns out that there is a bonding pattern for these acids and anions that should help you in drawing Lewis structures. There is always a central atom, less electronegative than O, which is surrounded by a combination of O atoms (anywhere from 0–3 of them) and OH groups (from 1–3 of them). See how this pattern is followed in the structures in Table 7.3.

Table 7.3 Lewis Structures of Oxoacids

$$:\overset{\displaystyle ||}{\underset{\displaystyle ..}{O}}$$
$$\overset{-}{:}\underset{..}{\overset{..}{O}}-\underset{+}{N}-\underset{..}{\overset{..}{O}}-H \longleftrightarrow \underset{..}{\overset{..}{O}}=\underset{+}{N}-\underset{..}{\overset{..}{O}}-H$$

Nitric acid

$$:\overset{\displaystyle ||}{O}:$$
$$H-\underset{..}{\overset{..}{O}}-C-\underset{..}{\overset{..}{O}}-H$$

Carbonic acid

$$\underset{..}{\overset{..}{O}}=N-\underset{..}{\overset{..}{O}}-H$$

Nitrous acid

$$\overset{\displaystyle H}{|}$$
$$:\overset{\displaystyle |}{O}:$$
$$H-\underset{..}{\overset{..}{O}}-P-\underset{..}{\overset{..}{O}}-H$$
$$:\overset{\displaystyle ||}{O}:$$

Phosphoric acid

$$:\overset{\displaystyle ||}{O}:$$
$$H-\underset{..}{\overset{..}{O}}-S-\underset{..}{\overset{..}{O}}-H$$
$$:\overset{\displaystyle ||}{O}:$$

Sulfuric acid

$$:\overset{\displaystyle ||}{O}:$$
$$H-\underset{..}{\overset{..}{O}}-S-\underset{..}{\overset{..}{O}}-H$$

Sulfurous acid

$$:\overset{\displaystyle ||}{O}:$$
$$H-\underset{..}{\overset{..}{O}}-Cl=\underset{..}{\overset{..}{O}}$$
$$:\overset{\displaystyle ||}{O}:$$

Perchloric acid

$$:\overset{\displaystyle ||}{O}:$$
$$H-\underset{..}{\overset{..}{O}}-\overset{..}{Cl}=\underset{..}{\overset{..}{O}}$$

Chloric acid

$$H-\underset{..}{\overset{..}{O}}-\overset{..}{Cl}=\underset{..}{\overset{..}{O}}$$

Chlorous acid

$$H-\underset{..}{\overset{..}{O}}-\overset{..}{\underset{..}{Cl}}:$$

Hypochlorous acid

Try drawing the Lewis structures for the oxoanions of the acids. In every case, you will be removing an H without its electron, so the oxygen will have a negative formal charge because it keeps the departed hydrogen's electron. How many resonance structures can you draw for each of theses anions? On the other hand, we can't draw any reasonable resonance structures for the oxoacids shown in Table 7.3 (except HNO_3). We conclude that the pi electrons are more delocalized in the anion than in the acid. If you remember that delocalization of pi electrons has a stabilizing effect, you should understand why these acids want to donate protons if they bump into an appropriate proton acceptor or base.

Molecular Geometry | *The shape of molecules*

One of the most important scientific discoveries of the 20th century was the elucidation of the structure of one molecule, DNA, the molecule that contains the blueprint for life. The knowledge about life's chemistry that has been acquired based on this structure fills hundreds of books and will likely lead to the cure for thousands of diseases. **Knowing the three-dimensional shape of molecules is essential to understanding the properties of molecules.**

Lewis structures of molecules and polyatomic ions show us that atoms are surrounded by covalent bonds and lone pairs. However, being restricted to two dimensions, Lewis structures cannot tell us how molecules fill three-dimensional space. Actually, the way in which atoms and lone pairs are arranged around atoms is fairly intuitive. Let's consider methane with four hydrogen atoms bonded to a central carbon. Figure 7.17 shows two different arrangements of the hydrogens around carbon. Can you guess which is the correct arrangement?

Figure 7.17 Two possible arrangements of hydrogen atoms around carbon in methane

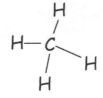

If you guessed the one on the right, you already have a good feel for molecular geometry. You might even think the other arrangement is unreasonable because the covalent bonds containing the negatively charged electrons are too close together. (Remember that covalent bonds contain negatively charged electrons.) As you would expect, the bonds repel each other such that the most stable arrangement is one that places the bonds as far apart from one another as possible. For methane, the geometry is called **tetrahedral,** which is based on the tetrahedron, a four-sided triangular pyramid. The farthest apart that four atoms bonded to a central atom can be is at the four corners of a tetrahedron, with the central atom in the center of the pyramid. In this arrangement, the angles between the four bonds are all about 109°, as shown in Figure 7.18. **Here's an important tip: Models are very helpful for visualizing molecules in three dimensions, and we encourage you to try building models of molecules. You can either buy a molecular model kit or you can make your own models with toothpicks and Styrofoam balls.**

Now let's consider the Lewis structures of ammonia and water, also shown in Figure 7.18. In ammonia, there are three covalent bonds and a lone pair; in water,

Figure 7.18 Four pairs of electrons around a central atom will be as far apart as possible, pointing to the corners of a four-sided tetrahedron.

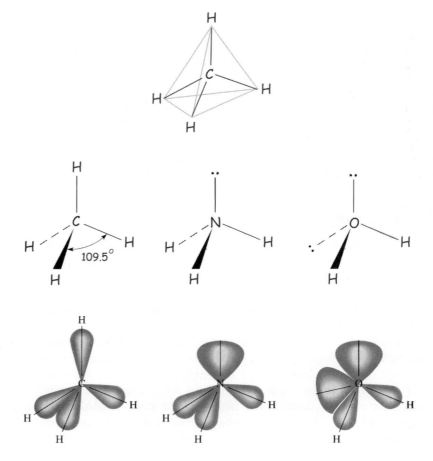

there are two bonds and two lone pairs. In both cases, the central atom is surrounded by four pairs of electrons although they are not all involved in bonding. As you would expect, electrons in lone pairs exert a repulsive force just like those in covalent bonds. So the four pairs of electrons (bonding and nonbonding) in ammonia and water want to be as far apart as possible; for four electron pairs, this would be a tetrahedral arrangement just like methane.

This approach to understanding molecular geometry is called the **valence-shell electron-pair repulsion (VSEPR)** model because it describes the arrangement of electron pairs around a central atom based on electrostatic repulsion between pairs of valence electrons. How many different types of arrangements are possible? There are only five. The type of arrangement depends on the number of electron pairs surrounding a central atom. As you have seen from drawing Lewis structures, there can be two, three, four, five, or six pairs of electrons (bonding or nonbonding) around a central atom. For each of these arrangements, the electron pairs are kept as far apart as possible. Table 7.4 shows the five basic geometries.

Most molecules are based on the trigonal planar and tetrahedral geometries. In fact, the billions of atoms in a DNA molecule are all centered within either a trigonal planar or tetrahedral geometry. An example of trigonal planar geometry is the carbonate ion, CO_3^{2-}. Because the atoms all lie in the same plane, its geometry is two dimensional and hence easily represented on paper, as shown in Figure 7.19. Note that carbonate has a double bond, but the extra bond doesn't affect the geometry because the pi bond is simply an additional connection to atoms already linked by a sigma bond. Even though the pi electrons in carbonate are delocalized over the entire ion, they still have no effect on the geometry; pi bonds are not taken into account when we determine molecular geometry—all that matters is the number of bonded atoms and lone pairs around a central atom.

What about molecules that contain more than one "central" atom? You can use the VSEPR model to predict the geometry around any atom bonded to more than one atom in a molecule or polyatomic ion. Let's consider the geometry of carbonic acid (H_2CO_3), which is the carbonate ion with two protons. With the additional hydrogens, we can consider the geometry around the oxygens that are bonded to hydrogen. Of course, the geometry around carbon is the same as in carbonate, which is trigonal planar. Do you notice anything familiar about the oxygens? They both look like the O atom in water except one of the hydrogens is replaced by carbon. Like in water, these oxygens are surrounded by two bonded atoms and two lone pairs. So the arrangement of the electron pairs around them is tetrahedral. Figure 7.19 also shows what carbonic acid would look like with the respective geometries around C and O.

Figure 7.20 shows some common molecules containing C, H, O, and N atoms. (Recall we mentioned in Chapter 2 that most of the molecules in the world are comprised of these four atoms.) Take a minute and try to identify the geometry around each of the atoms bonded to more than one other atom. This turns out to be fairly

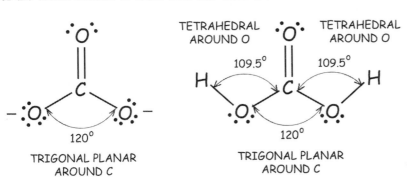

Figure 7.19 Geometric arrangement of atoms in carbonate ion and carbonic acid

Table 7.4 Arrangement of Electron Pairs About a Central Atom (A) in a Molecule and Geometry of Some Simple Molecules and Ions in Which the Central Atom Has No Lone Pairs

Number of Electron Pairs	Arrangement of Electron Pairs	Molecular Geometry	Examples
2	Linear (180°)	B—A—B Linear	$BeCl_2$, $HgCl_2$
3	Trigonal planar (120°)	Trigonal planar	BF_3
4	Tetrahedral (109.5°)	Tetrahedral	CH_4, NH_4^+
5	Trigonal bipyramidal (90°, 120°)	Trigonal bipyramidal	PCl_5
6	Octahedral (90°)	Octahedral	SF_6

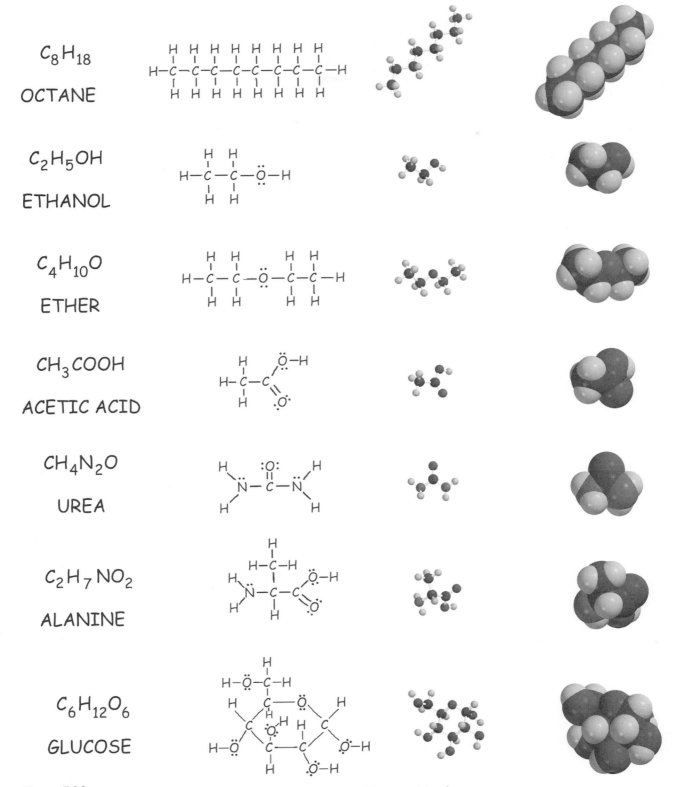

Figure 7.20 Lewis structures, ball-and-stick models, and space-filling models of some common molecules

straightforward if you think of all of these molecules as having architecture built from units derived from methane, ammonia, water, and carbonic acid.

Let's start with octane (C_8H_{18}), which is a component of gasoline. It looks like it was made by stringing together eight methane molecules. The difference in the carbon bonding in octane relative to methane is that each carbon is bonded to either three hydrogens and one carbon (on the ends) or two hydrogens and two carbons between the ends. Every carbon will therefore have the same tetrahedral geometry as the carbon in methane. By the way, the CH_3 group at the ends of octane is called a methyl group. The ethanol molecule, the active ingredient in beer, wine, and booze, looks like methane on the left and water on the right, except the O is bonded to hydrogen and carbon instead of two hydrogens. Thus, the arrangement of electron pairs around the two carbons and the oxygen is tetrahedral. The CH_3CH_2 group in ethanol is called an ethyl group, and this molecule is often referred to as ethyl alcohol. Diethyl ether, an anesthetic that was commonly used to knock out patients before surgery, can be thought of as a water molecule in which the hydrogens have been replaced by two ethyl groups.

Now look at acetic acid, the acid in vinegar. It looks like carbonic acid where one of the OH groups (called hydroxyl groups) is replaced by a methyl group. Like carbonic acid, the carbon with the double bond is trigonal planar; the arrangements of electron pairs around the other carbon and the oxygen are tetrahedral. Earlier, we showed the formula of acetic acid as $HC_2H_3O_2$ (see p. 36). From now on, we will write its formula as CH_3COOH because it more correctly represents the bonding between the atoms.

The molecule urea is a major component in urine and other bodily fluids. With its trigonal planar carbon, it looks like carbonic acid in which the two OH groups have been replaced by NH_2 groups (which are often called *amino groups* because they're derived from ammonia). The arrangement of electron pairs around each nitrogen is tetrahedral just like that in ammonia because the bonding is essentially the same except a hydrogen is replaced by a carbon. The amino acid alanine (can you guess why it's called an amino acid?) is one of the 20 common amino acids that serve as the building blocks of proteins. You should be able to see the basic units derived from methane, ammonia, water, and carbonic acid in this molecule and identify the corresponding geometries.

The last molecule in Figure 7.20 is the sugar glucose, which has a ring structure. What are the geometries around all of the atoms bonded to more than one atom in this molecule? All of the carbons are tetrahedral like that in methane, and all of the oxygens are like the oxygen in water.

Note the models of the compounds in Figure 7.20 do not show the lone pairs, they just show the atoms oriented in a way that is predicted by the VSEPR model. If we depict the structure of water and ammonia without the electron pairs, we see the actual shape of the molecules are those shown in Figure 7.21. The bond angles in ammonia and water are all about 109° because this places the four electron pairs farthest apart, but the geometries of the two molecules are different. Ammonia is shaped like a pyramid, and water is bent.

In addition to the five basic geometries, there are several other geometries possible depending on the number of lone pairs surrounding a central atom. Table 7.5 shows all of the possible shapes when lone pairs are present. Because most molecules have either trigonal planar or tetrahedral electron arrangements, we will focus only on examples with geometries related to these basic arrangements. These geometries are: tetrahedral, trigonal planar, pyramidal, or bent. Note that if there are no lone pairs, the geometry is the same as the arrangement of electron pairs. So the shape of methane is tetrahedral.

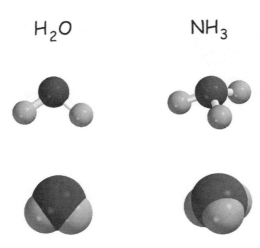

H_2O NH_3

The following steps summarize our approach to determining molecular geometry.

1. Draw the Lewis structure of the molecule.
2. Determine the number and hence the arrangement of electron pairs around the central atom.
3. If the central atom does not possess lone pairs, then the geometry of the molecule, which is determined only by the positions of the atoms, is the same as the electron arrangement.
4. If the central atom possesses one or more lone pairs, then the geometry of the molecule corresponds only to the portion of the electron arrangement involving bonding pairs.

7.6

EXAMPLE

Indicate the geometry for each of the following molecules. (a) H_2S, (b) AsH_3, and (c) BF_3.

Answer *(a) The Lewis structure of H_2S is*

$$H - \overset{..}{\underset{..}{S}} - H$$

Because there are two lone pairs on the S atom, the H_2S molecule has a bent geometry, like that of H_2O.
(b) The Lewis structure of AsH_3 is

$$H - \overset{..}{As} - H$$
$$|$$
$$H$$

This molecule has three bonding pairs and one lone pair, a combination similar to that of ammonia. Therefore, the geometry of AsH_3 is trigonal pyramidal, like NH_3.
(c) The Lewis structure of BF_3 is

$$F$$
$$|$$
$$F - B - F$$

There are no lone pairs on the B atom. (BF_3 is an example of the incomplete octet discussed earlier. For simplicity, we omit the lone pairs on F atoms.) Thus the geometry of the molecule is the same as the arrangement of the electron pairs, trigonal planar.
    (Check the website for more problems on molecular geometry.)

Table 7.5   Geometry of Some Simple Molecules and Ions in Which the Central Atom Has One or More Lone Pairs

Class of Molecule	Total Number of Electron Pairs	Number of Bonding Pairs	Number of Lone Pairs	Arrangement of Electron Pairs	Geometry	Examples
AB_2E	3	2	1	Trigonal planar	Bent	SO_2
AB_3E	4	3	1	Tetrahedral	Trigonal pyramidal	NH_3
AB_2E_2	4	2	2	Tetrahedral	Bent	H_2O
AB_4E	5	4	1	Trigonal bipyramidal	Distorted tetrahedron (or seesaw)	SF_4
AB_3E_2	5	3	2	Trigonal bipyramidal	T-shaped	ClF_3
AB_2E_3	5	2	3	Trigonal bipyramidal	Linear	I_3^-
AB_5E	6	5	1	Octahedral	Square pyramidal	BrF_5
AB_4E_2	6	4	2	Octahedral	Square planar	XeF_4

Predicting Polarity Based on Molecular Geometry

In Chapter 2 you saw that molecules can be classified as either polar or nonpolar. By now you should be able to better understand the basis of polarity in molecules. If two atoms joined in a covalent bond have different electronegativities, the more electronegative atom pulls the electrons toward itself. In a diatomic molecule like HF, the result is a molecule that is slightly negative on the F end and slightly positive on the H end.

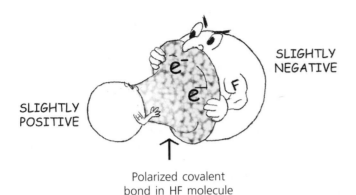

Polarized covalent
bond in HF molecule

We call a polar molecule like HF a dipole because it has two poles—a positive pole and a negative pole. These poles are analogous to the north and south poles of a bar magnet. Using electronegativities, we can assess the degree of polarity (that is, how much the electron density is shifted) in a molecule, and knowing the polarity of a molecule helps us understand and predict molecular behavior.

It should make sense to you that the degree of polarity in a bond depends on the amount of charge at the ends of the molecule—the bigger the charge, the more polar the bond. The charge at both ends would have the same magnitude, but one end would be positive and the other negative. The charge would also be less than one; if they were +1 and −1, they'd become a cation and an anion. The degree of polarity also depends on the distance between the charges (that is, the distance between the nuclei of the atoms). A convenient measure of the polarity of a molecule is called the **dipole moment,** and it is simply the product of the charge times the distance. Dipole moments have been measured for some molecules, and they are expressed in debye

THE DIPOLE MOMENT IS A MEASURE OF THE POLARITY OF A BOND. IT IS MEASURED IN DEBYE UNITS, NAMED AFTER ME.

Table 7.6 Dipole Moments of Some Polar Molecules		
Molecule	Dipole Moment (D)	Electronegativity Difference
HF	1.92	1.9
HCl	1.08	1.1
HBr	0.78	0.7
HI	0.38	0.4
H_2O	1.87	1.4
H_2S	1.10	0.4
NH_3	1.46	0.9
SO_2	1.60	1.0

(D) units, named after the Dutch-American chemist and physicist **Peter Debye.** For example, the dipole moment of HF, where the electronegativity difference between H and F is 1.9 (see Figure 7.2), is 1.92 D. As you can see in Table 7.6, there is a strong correlation between electronegativity difference and dipole moment.

Dipole moments are represented by arrows with the arrowhead pointing toward the more electronegative atom, as shown in Figure 7.22, which makes it easy to represent the dipole moments in diatomic molecules. Of course, diatomic molecules containing two atoms of the same element have no difference in electronegativity (that is, the electrons are shared equally), so there is no dipole moment and they are nonpolar.

But how do we predict the polarity of a polyatomic molecule? To do so, we need to know the geometry of the molecule. Consider the water molecule with its bent geometry. Because oxygen is more electronegative than H, electrons will be pulled toward O. In Figure 7.22, the electron pull in each O—H bond is indicated by an arrow with the arrowhead pointing toward the O. The overall direction of the electron

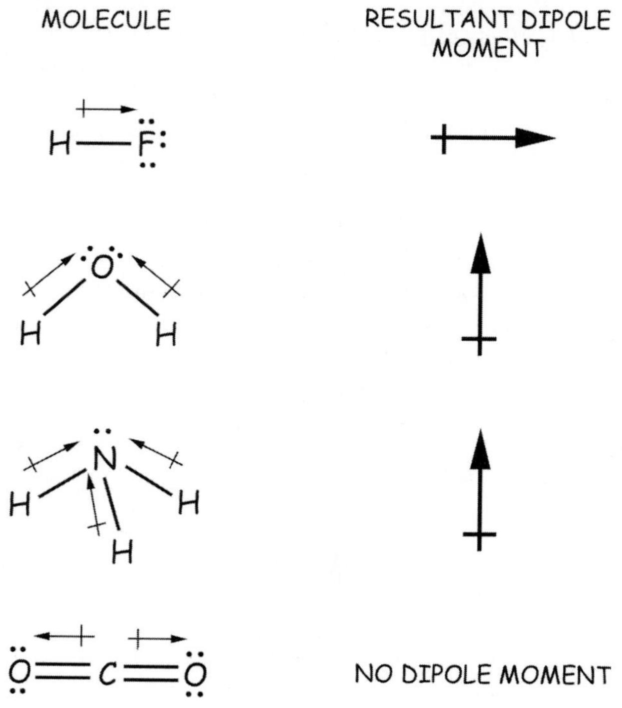

Figure 7.22 Bond moments and resultant dipole moments for some common molecules

pull or the shift in electron density would be represented by an arrow pointing straight up. Mathematically, the overall dipole moment is the vector sum of the dipole moments of the individual bonds.

This approach will also tell us that a molecule can be nonpolar even if the individual bonds are polar. Consider the linear carbon dioxide (CO_2) molecule ($O=C=O$). (Why is CO_2 linear?) Because oxygen is more electronegative than carbon, electrons will be pulled toward O, as shown in Figure 7.22. The electron pull in each $C=O$ bond is the same in magnitude but opposite in direction so they cancel each other out. Thus, the overall dipole moment is zero, making CO_2 a nonpolar molecule.

Predict whether the following compounds possess a dipole moment: (a) NF_3, (b) CH_4, (c) H_2S, (d) CS_2, (e) BF_3, (f) CH_2Cl_2.

Answer *(a) NF_3 has a trigonal pyramidal shape, and the N atom has a lone pair. The electron density is pulled toward the more electronegative F atom, and the resultant dipole moment is along the lone pair. Therefore, NF_3 is a polar molecule.*

$$F \quad \ddot{N} \quad F$$

resultant dipole moment

(b) The C atom is only slightly more electronegative than H. Because of the tetrahedral geometry of CH_4, the four bond moments exactly cancel one another. Thus, CH_4 is a nonpolar molecule.

$$\begin{array}{c} H \\ H \quad C \quad H \\ H \end{array}$$

(c) H_2S has a bent shape just like water. Therefore, the two bond moments give rise to a resultant dipole moment and the molecule is polar.

resultant dipole moment

$$H \qquad H$$

(d) CS_2 has a linear shape just like CO_2. Therefore, although S is more electronegative than C, the two bond moments cancel each other, making it a nonpolar molecule.

$$S=C=S$$

(e) The geometry of BF_3 was shown to be trigonal planar in Example 7.6. Because of its symmetrical shape, the three B—F bond moments cancel one another, and the molecule is nonpolar.

$$\begin{array}{c} F \\ B \\ F \qquad F \end{array}$$

—Continued next page

Continued—

(f) CH_2Cl_2 has a tetrahedral geometry similar to that of CH_4. However, because the substituents (end atoms) are not all identical, the bond moments do not cancel one another. Consequently, the molecule has a resultant dipole moment, and it is polar.

$$
\begin{array}{c}
H \\
\updownarrow \\
H \rightarrow C \rightarrow Cl \\
Cl
\end{array}
$$

resultant dipole moment

(Check the website for more problems on polarity of molecules.)

Determining experimentally whether or not a molecule has a dipole moment is the best way to predict if a molecule is polar or nonpolar. However, a general rule is that most hydrocarbons (that is, compounds containing only C and H atoms) are nonpolar. Knowing whether a molecule is polar or nonpolar has a lot of practical applications based on solubility. For example, the extraction of molecules with medicinal properties from plants requires the systematic use of solvents with different polarities. At the molecular level, the basis for solubility is the nature of the forces that hold molecules together, collectively called *intermolecular forces.*

Intermolecular Forces | *The glue between molecules*

Now that you understand the nature of the bonds between atoms, we will look at the forces that hold molecules together in the liquid or solid state. In order for substances to exist as a liquid or a solid, the molecules must stick to one another. This stickiness is what we call **intermolecular forces** or **IMFs.** You already have some practical experience with IMFs if you have ever touched maple syrup, glue, or Scotch tape. These substances owe their stickiness at the macroscopic level to the microscopic IMFs acting between molecules. (Knowing chemistry enables you to understand all sorts of commonplace phenomena at the most fundamental level.)

All IMFs are electrostatic forces, like the attractive forces between two ions. To understand how this works, let's consider the interaction between an ion and a polar molecule. A familiar example of this is when NaCl dissolves in water. As you already know from Chapter 2, water molecules surround dissociated Na^+ and Cl^- ions in a process called hydration. But why do the water molecules surround these ions? There must be a stabilizing force to compensate for breaking the ionic bonds in NaCl. The stabilizing force is essentially the same as that in an ionic bond: positive attracts negative. Because water has a dipole moment, it has a positive end (the hydrogens) and a negative end (the oxygen). In an aqueous NaCl solution, water molecules arrange themselves around Na^+ ions such that their negative ends point to the ion and are attracted by an electrostatic force. This attraction is not as strong as that in an ionic bond because the oxygen end of the molecule has only a partial charge (that is, it's less than -1). Conversely, the water molecules surrounding Cl^- ions point their positive ends toward the negative ion. These attractive interactions are called **ion-dipole forces,** which are the type of IMF that exist between all ions and water molecules in an aqueous solution (Figure 7.23).

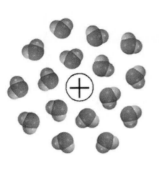 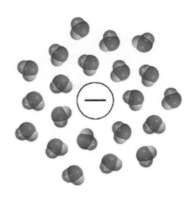

By the same token, we can see how polar molecules can stick together among themselves because the attractive forces here are also electrostatic in origin. Consider the HCl molecule with a boiling point of $-85°C$. That's pretty cold. At temperatures below the boiling point, HCl will exist as a liquid due to attractive forces between the molecules. In fact, boiling point is a measure of the amount of energy needed to break the IMFs in a substance. At $-85°C$, the HCl molecules have enough thermal energy to overcome the IMF. Below $-85°C$, most of the molecules remain in the liquid state and align themselves as shown in Figure 7.24. This type of IMF is called a **dipole-dipole force,** and it occurs between all polar molecules. Like ion-dipole forces, it is simply a case of positive attracts negative. As you might guess, dipole-dipole forces are weaker than ion-dipole forces, which, in turn, are much weaker than ionic and covalent bonds (see Table 5.2). Indeed, the low boiling point of HCl is evidence that these forces are relatively weak.

There is, however, a special type of dipole-dipole force that is unusually strong. It occurs in many substances—water, ammonia, alcohols, acids, and most of the molecules in your body. This extra-strong dipole-dipole force is called **hydrogen bonding,** and, as the name implies, it involves hydrogen. It turns out that molecules containing H covalently bonded to N, O, or F have considerably higher boiling points compared with other polar molecules. Compare the boiling points for the simple hydrogen-containing compounds in Table 7.7. Note that they are arranged according to group numbers so you can easily compare the differences between similar compounds. For the Group 4A elements, we see that the compound containing Si (silane, SiH_4) has a higher boiling point than the one containing C (methane, CH_4). Therefore, the IMFs of the former are stronger. As we will describe, the boiling points of similar compounds tend to increase as the size and mass of the molecule increase. According to this trend, we would expect that all of the compounds containing second-period elements in Groups 5A–7A would have lower boiling points than their third-period counterparts. Instead, we see that the boiling points (and therefore the magnitude of the IMFs) of the compounds containing N, O, or F bonded to H are unusually high. There are countless examples of other molecules containing N—H, O—H, or F—H bonds with unusually high boiling points (as well as other properties that depend on the strength of IMFs). In particular, complex biological molecules owe their stability and many of their

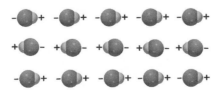

Figure 7.24 HCl molecules interact with each other through dipole-dipole forces.

Table 7.7 Boiling Points of Hydrogen-Containing Compounds of the Group 4A, 5A, 6A, and 7A Elements

Group	Compound	B.P. (°C)
4A	CH_4	−161
	SiH_4	−111
5A	NH_3	−33
	PH_3	−88
6A	H_2O	100
	H_2S	−61
7A	HF	20
	HCl	−85

biochemical properties to hydrogen bonding. Perhaps the most well known example is the hydrogen bonding that holds the double helix together in DNA.

Because the force is so strong between molecules containing hydrogen bonded to N, O, or F, and because such molecules are prevalent, they have been given a special name. But don't let the name hydrogen *bond* confuse you; hydrogen bonds are a type of IMF, different and weaker than ionic and covalent bonds. You can think of them as really strong dipole-dipole forces. Although the force is electrostatic, the importance of hydrogen bonding has led to the use of dotted lines to indicate their presence, as shown in figure 7.25.

The hydrogen bonding arrangement for water molecules in ice is shown in Figure 7.26. Each water molecule forms four hydrogen bonds with four other water molecules. This is a very stable structure, but there is a void space. Why does ice float in water? Water is the only common substance whose liquid state is denser than its solid state. (And if it weren't for this property, life as we know it would not exist.) Imagine the disruption of some hydrogen bonds during melting, which frees some water molecules from the three-dimensional network. These freed molecules can fit in the holes in the lattice resulting in more molecules per unit volume; it becomes more dense.

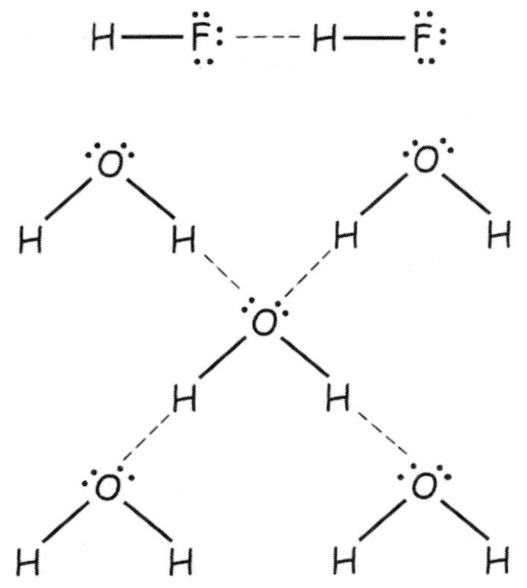

Figure 7.25 Hydrogen bonding in HF and H_2O

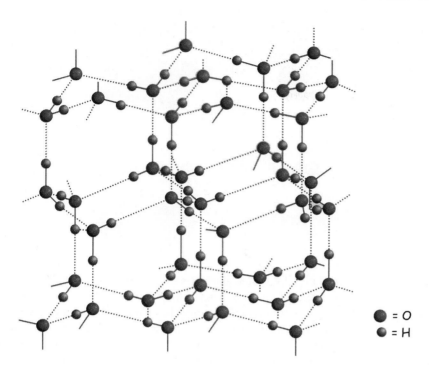

Figure 7.26 Hydrogen bonding in ice

● = O
● = H

Why is the hydrogen bond so strong? One contributing factor is the exposed nucleus (a proton) of the hydrogen atom when it's bonded to N, O, or F. With its only electron pulled toward the other end of the molecule, the hydrogen end is essentially a naked proton. The effective charge of the H is substantially greater than it would be on other nonmetals with complete inner shells of electrons shielding the nuclear charge.

The Rest of the IMF Story

So far, the discussion of IMFs has been pretty straightforward: positive attracts negative. The situation gets more complicated when we try to understand the way nonpolar molecules interact. If there is no dipole moment, what is there for one nonpolar molecule to attract another nonpolar molecule? For example, what holds N_2 molecules together in liquid nitrogen? There must be some attractive force because nonpolar substances also exist as solids and liquids. Table 7.8 shows the boiling points of some simple nonpolar molecules. As we saw earlier, there is a clear trend: the strength of the IMFs increases with the size of molecules and number of electrons. What could be the nature of such a force?

The origin of the IMFs that exists between nonpolar molecules is based on the ability of these molecules to create a *temporary* dipole in another nonpolar molecule. To understand how this works, let's first see how a temporary dipole can be created in a nonpolar molecule. Imagine a sodium ion approaching an I_2 molecule as shown in Figure 7.27. When Na^+ is near the end of the nonpolar iodine molecule, the valence electrons of nearby I would be attracted to the positive ion. The result is an unequal distribution of electrons leading to a dipole moment in I_2. We call this an **induced dipole,** and we say that the sodium ion induces a dipole in iodine. The word *induce* is appropriate because it means to "move by persuasion or influence." You can also imagine that a polar molecule with its positive end pointing toward a nonpolar molecule would also induce a dipole in the nonpolar molecule.

So how does this apply to nonpolar molecules when there are no ions or polar molecules around? According to the accepted model, nonpolar molecules can have

Table 7.8 Boiling Points of Simple Nonpolar Molecules

Molecule	B.P. (°C)
H_2	−253
N_2	−196
O_2	−183
F_2	−145
Cl_2	−34
Br_2	59
I_2	183

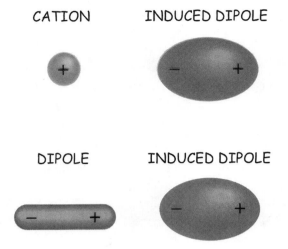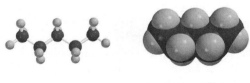

Figure 7.27 A dipole can be induced in a nonpolar molecule by the proximity of a charged species.

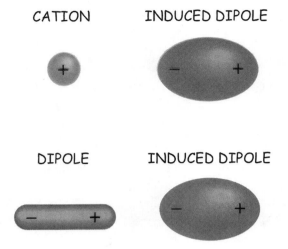

CATION

INDUCED DIPOLE

DIPOLE

INDUCED DIPOLE

temporary dipoles, which, in turn, induce dipoles in neighboring molecules. There are two ways that a molecule could become temporarily polarized. Because electrons are always moving around, at some instant there may be a shift in electron density to one end of a nonpolar molecule, causing a temporary dipole. A collision between two non-polar molecules could also cause a momentary shift in electron density. In either case, the temporary dipole could induce a dipole in a nearby molecule, and the induced dipole would interact electrostatically with the original temporary dipole. Of course, other nearby molecules would also be affected similarly, and eventually instantaneous dipoles are dispersed throughout all the molecules. But collisions between molecules are constantly destroying old temporary dipoles and creating new ones. This type of IMF is called a **dispersion force.** The strength of dispersion forces depends on size, as shown in Table 7.8. Bigger molecules have more electrons, and their electron density is more easily distorted, leading to larger temporary dipoles.

The strength of dispersion forces also depends on the shape of molecules. Comparison of boiling points for molecules with the same molar mass, but different shapes, indicates the amount of surface area that can come in contact with other molecules is important. As shown in Figure 7.28, the molecule with the more compact shape, neopentane (C_5H_{12}), has the lower boiling point, and the one with the elongated shape (and greatest surface area), *n*-pentane (also C_5H_{12}), has the higher boiling point.

Figure 7.28 Structures and boiling points of nonpolar hydrocarbons with the same mass

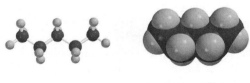

n-pentane: b.p. = 36.1 °C

neopentane: b.p. = 9.5°C

The dependence of boiling point on size and shape applies to polar molecules as well. This is because all molecules, polar and nonpolar alike, have dispersion forces, which can be fairly strong if the molecule is large.

Let's summarize the important points about IMFs:

- IMFs are electrostatic interactions between molecules: positive attracting negative.
- IMFs are weaker than ionic bonds and covalent bonds.
- IMFs must be broken to melt and boil compounds.
- Ion-dipole forces occur when ionic compounds dissolve in water.
- Dipole-dipole forces occur between polar molecules.
- Hydrogen bonds are especially strong dipole-dipole forces that occur between polar molecules containing N—H, O—H, or F—H bonds.
- Dispersion forces occur in all compounds, but they are the only IMF between nonpolar molecules.

7.8

EXAMPLE

*Predict the types of intermolecular forces present in the following compound:
(a) HBr, (b) CS_2, and (c) CH_3OH.*

Answer   *(a) HBr is a polar molecule. Therefore, the forces present are dipole-dipole and dispersion forces. (b) CS_2 is a nonpolar molecule. The only forces present are dispersion forces. (c) The predominant forces present are due to hydrogen bonding. There are also dispersion forces.*
   (Check the website for more problems on IMFs.)

Solubility and IMFs

Now that you know about IMFs, you can better understand solubility. You already know how ionic compounds dissolve in water. The ion-ion forces between the cations and anions in the solid are replaced by ion-dipole forces between water and the ions. What happens when polar molecules are dissolved in water? A familiar example is the dissolution of sugar in water. Figure 7.20 shows the structure of the sugar glucose, which is very soluble in water. (Table sugar, sucrose, looks like two glucose molecules bonded together and is also very soluble in water.) What kinds of IMFs contribute to the solubility of glucose in water? There are five OH groups in glucose, all capable of hydrogen bonding. In the solid state, these hydrogen bonds require a lot of energy to break as evidenced by glucose's high m.p. of about 150°C (compare with Br_2, which has a similar mass but only a m.p. of −7°C because it is nonpolar). When glucose is added to water, the hydrogen bonds between the glucose molecules are replaced by equally strong hydrogen bonds with water molecules. This interaction enables water molecules to surround the individual sugar molecules and keep them in solution.

The dissolution of both ionic compounds and molecules like sugars in water requires that water molecules surround the dissolved species. As such, there is a limit to their solubility, which is why you can't dissolve a cup of salt or sugar in a teaspoon of water. There are, however, some compounds that are completely soluble in water. Such compounds in water are said to be **miscible,** which means they and water are soluble in each other in all proportions. An example is ethanol, which looks like water where a hydrogen atom is replaced by an ethyl group, as shown in Figure 7.20. Ethanol is miscible with water because a drop of ethanol will dissolve in a liter of water, a drop of water will dissolve in a liter of ethanol, and every proportion in between is completely soluble—they're soluble in all proportions. Like water, ethanol can form hydrogen bonds and easily replaces water molecules in its network of hydrogen bonds. Similarly, water can replace ethanol molecules in liquid ethanol.

Nonpolar compounds are generally very soluble in nonpolar solvents because they all stick together with the same IMFs—dispersion forces. The word *solvent* is sometimes used to describe smelly and nasty liquids used as cleaning fluids such as turpentine, which are often nonpolar liquids. One reason that their odor is so noticeable is that these nonpolar compounds have only weak dispersion forces holding the molecules in the liquid state. As a result, there is a high rate of escape from the liquid state, and many solvent molecules will find the sensory molecules in your nose. The reason nonpolar solvents are good for tough cleaning jobs is that usually the easy-to-clean dirt is water soluble and the hard-to-clean dirt is nonpolar oil and grease.

7.9

EXAMPLE

Predict whether the following compounds would be soluble in water: (a) KBr, (b) CCl₄, and (c) CH₃OCH₃.

Answer   *(a) KBr is an ionic compound. The K^+ and Br^- ions will be hydrated in solution. Therefore, it is soluble in water. (b) CCl₄ (carbon tetrachloride) is a nonpolar molecule. There are only weak dipole-induced dipole and dispersion forces between CCl₄ and H₂O. Thus, CCl₄ is not soluble in water. (c) CH₃OCH₃ (dimethyl ether) can form hydrogen bonds with water molecules. Therefore, it should be (and indeed it is) soluble in water.*

(Check the website for more problems on solubility.)

Test your understanding of the material in this chapter

IMPORTANT TERMS

Explain the following terms in your own words:

dipole-dipole forces, p. 139
dipole moment, p. 135
dispersion forces, p. 142
electronegativity, p. 106
expanded octet, p. 125
formal charge, p. 122
hydrogen bonding, p. 139
induced dipole, p. 141

intermolecular forces (IMFs), p. 138
ion-dipole forces, p. 138
Lewis structure, p. 110
lone pair, p. 110
miscible, p. 143
molecular orbital theory, p. 111

octet rule, p. 116
polar molecule, p. 105
resonance, p. 119
resonance structure, p. 119
valence-shell electron-pair repulsion (VSEPR), p. 129

UNDERSTANDING CHEMISTRY

Summarizing Problem
The molecule benzene has the chemical formula C_6H_6 and has a ring structure.

(a) Draw the Lewis structure for benzene including any resonance structures that obey the octet rule.
(b) Predict the geometry around all of the carbon atoms.
(c) Describe the overall shape of the molecule.
(d) Describe the molecular orbitals in benzene.
(e) Predict whether or not the following compounds would be soluble in benzene: (1) NaCl, (2) CH₃OH, and (3) C₈H₁₈.

Answers: (a) The Lewis structure of benzene is

$$
\begin{array}{c}
\text{H} \\
\vert \\
\text{C}
\end{array}
$$

A simpler way of drawing the structure of the benzene molecule and other compounds containing the "benzene ring" is to show only the skeleton and not the carbon and hydrogen atoms. By this convention, the resonance structures are represented by

Note that the C atoms at the corners of the hexagon and the H atoms are all omitted, although they are understood to exist. Only the bonds between the C atoms are shown.

Remember this important rule for drawing resonance structures: *The positions of electrons, but not those of atoms, can be rearranged in different resonance structures.* In other words, the same atoms must be bonded to one another in all the resonance structures for a given species.

(b) If we ignore the pi bond, then each C atom has three bonding pairs. Therefore, it has a trigonal planar geometry.

(c) The overall geometry is planar.

(d) Each C atom forms three sigma bonds with two C atoms and one H atom. In addition, the pi bonds between the C atoms are delocalized over the entire molecule (Figure 7.29).

(e) (1) NaCl is an ionic compound and is not soluble in benzene. The ion-induced dipole forces are too weak. (2) CH_3OH is a polar molecule capable of forming hydrogen bonds. It is not soluble in benzene because the dipole-induced dipole forces are too weak. (3) C_8H_{18} (octane) is a nonpolar molecule. As a result, the strong dispersion forces between octance and benzene make C_8H_{18} soluble in C_6H_6.

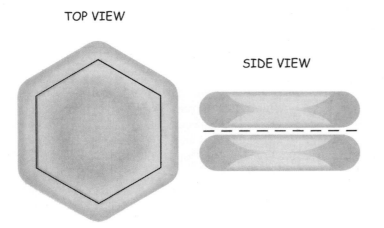

TOP VIEW

SIDE VIEW

Figure 7.29   Delocalized pi orbitals in benzene

8 Chemical Equilibrium
Back and forth, back and forth

I n this chapter you will build on your knowledge of chemical reactions (Chapters 3 and 4), energy changes (Chapter 5), and chemical bonding (Chapter 7) to learn one of the most important concepts in chemistry— **chemical equilibrium.** We introduced you to chemical equilibrium in Chapter 3 when we described the behavior of acetic acid in water (p. 36). Recall that we used a double arrow ($\rightleftharpoons$) in the chemical equation to show that the reaction is reversible.

$$CH_3COOH(aq) + H_2O(l) \rightleftharpoons CH_3COO^-(aq) + H_3O^+(aq)$$

The double arrows mean that both the forward and reverse reactions are occurring simultaneously. As such, it actually describes two chemical reactions:

$$CH_3COOH(aq) + H_2O(l) \longrightarrow CH_3COO^-(aq) + H_3O^+(aq)$$
$$CH_3COO^-(aq) + H_3O^+(aq) \longrightarrow CH_3COOH(aq) + H_2O(l)$$

Reversible Chemical Reactions

Most chemical reactions are reversible, at least to some extent. At the start of a reversible process, the reaction proceeds toward the formation of products. As soon as some product molecules are formed, the reverse process begins to take place, and reactant molecules are formed from product molecules. Chemical equilibrium is reached when the rates of both the forward and reverse reactions are equal, that is, when the amount of product formed per unit time is the same as the amount of reactant formed by the reverse reaction per unit time (that is, every second or every minute). When a system is at equilibrium, the amounts of all species involved in the reaction remain constant. **But it is very important to remember that although there is no net change in the amounts of reacting species, molecules are constantly reacting.** Chemical equilibrium is a dynamic process because molecules are being formed and being consumed at every instant. Because the opposing processes are happening at the same rate, the amounts of the reactants and products never change.

The defining feature of chemical equilibrium is that the rates of two opposing processes are equal. Therefore, to understand chemical equilibrium, you must first know how the rates become

equal. Let's consider a macroscopic example of two opposing processes reaching a dynamic equilibrium. Imagine a room filled with thousands of bumblebees randomly flying around. Adjacent to the room is an identical room, and the wall separating the two rooms has a very small window. We'll call them rooms 1 and 2, where room 1 contains all the bees in the beginning. When the window is opened, the bees can pass from room 1 into room 2 (the forward reaction) and back again (the reverse reaction).

$$\text{bees in room 1} \rightleftharpoons \text{bees in room 2}$$

The rate of bees going into room 2 is dependent on the number of bees in room 1. If there were 10,000 bees, the number of bees moving through the window per unit time would be about 10 times faster than if there were only 1000 bees. Of course, the rate of bees moving from room 2 → room 1 is initially zero because there were no bees in the empty room.

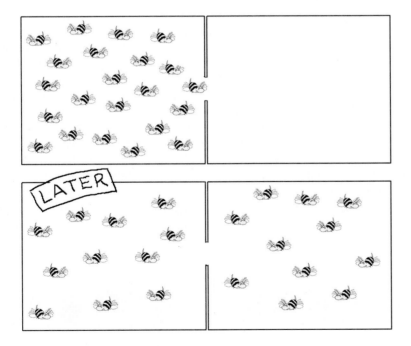

So initially the rate of bee movement from room 1 → room 2 is much greater than the rate of bee movement from room 2 → room 1. As time passes, the number of bees in room 2 increases, and some of these bees will move back into room 1; as the number of bees increases in room 2, the rate of bee movement into room 1 also increases. Meanwhile, the rate of bee movement from room 1 to room 2 is decreasing because the number of bees is decreasing. As a result, the faster rate for 1 → 2 is decreasing, and the slower rate for 2 → 1 is increasing. The two rates will eventually be equal. When the rates are equal, there can be no net change in the number of bees in each room. But there is constant movement—bees coming and going all the time. In this particular case, the rates will be equal when the numbers of bees in both rooms are the same because the rooms are identical. If the rooms were different sizes, this would not be the case. This analogy cannot be taken too far, however, because the amounts of reactants and products are rarely equal when a chemical reaction reaches equilibrium.

If you understand the bee analogy, you have grasped the essential ideas of chemical equilibrium. Let's look at a simple example of a reversible physical process—the equilibrium between liquid water and water vapor:

$$H_2O(l) \rightleftharpoons H_2O(g)$$

Imagine the following situation in which some liquid water is placed in a closed container. Initially the space above the water is empty. (This condition can be created if a barrier was first used to cover the water and all the air and water molecules above it are pumped away.) The water molecules in the liquid state are constantly colliding with one another. Every now and then a molecule will gain sufficient kinetic energy to break away from the intermolecular forces (mostly hydrogen bonding) to enter the empty space. Soon there will be enough molecules in the space above the liquid, and a vapor phase is established. The process is called *vaporization* or *evaporation*. The rate at which water molecules leave the liquid (that is, the rate of evaporation) depends on the surface area of the liquid and the temperature. A larger area will allow more molecules to escape, and a higher temperature will increase the average kinetic energy of the molecules so the rate will also increase. Because the container is closed, the molecules in the vapor will eventually collide with the water surface, and in so doing, some of the molecules will become trapped in the liquid state. The rate of molecules returning to the liquid state (a process called *condensation*) will depend on the number of water molecules in the vapor and the surface area of the water.

As with the bumblebees, the rate of evaporation at the beginning of the process is much greater than the rate of condensation because there aren't many water molecules in the vapor phase. As time goes on, however, the concentration of water molecules in the vapor phase increases and, hence, so does the rate of condensation. Eventually the rate of evaporation becomes equal to the rate of condensation, and the system reaches a dynamic equilibrium. Under this condition, the pressure exerted by the molecules in the vapor state is called the *equilibrium vapor pressure,* or simply *vapor pressure.* Because the system is at equilibrium, the vapor pressure does not change with time even though there is much back-and-forth molecular movement

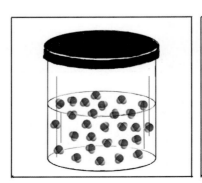
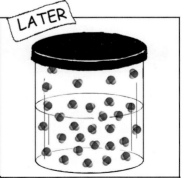

between the liquid and the vapor states. As you might expect, the vapor pressure of water (or any other liquid) increases with temperature. On the other hand, the surface area of the water does not affect the magnitude of the vapor pressure. The reason is that a larger surface area allows more molecules to escape from the liquid state, but it also lets more molecules return to the liquid state, so the two effects balance out.

The bumblebees and the liquid-gas equilibrium for water are useful ways to understand three fundamental features of chemical equilibrium:

1. Molecules are always reacting.
2. The rate of a reaction depends on the number of molecules.
3. Once the rates of opposing reactions are equal, there can be no net change despite the fact that reactions are occurring in both the forward and reverse directions.

Let's look at our example from Chapter 3 (p. 36)—the ionization of acetic acid in water.

$$CH_3COOH(aq) + H_2O(l) \rightleftharpoons CH_3COO^-(aq) + H_3O^+(aq)$$

Unlike the previous examples involving movement between two rooms or two phases, this process takes place in aqueous solution where all of the reacting species are mixed together in water (which is itself a reactant). The chemistry is straightforward: a proton is transferred from acetic acid to water in the forward reaction, and a proton is transferred from a hydronium ion to acetate ion in the reverse reaction. In order for a reaction to occur, molecules must collide, so the rate of the reaction depends on the concentration of molecules. It's analogous to the collision between blindfolded swimmers in a swimming pool—the rate of contact increases as the number of swimmers increases.

It's easy to understand the swimming pool analogy because we can envision people bumping into each other in a pool. Likewise, it's important to imagine molecules colliding with one another in solution to understand how they react. To help you imagine the reaction of acetic acid and water, let's choose a specific concentration of acetic acid and determine the relative amounts of acetic acid and water molecules present. Suppose we have 1 liter of a 1 M solution of acetic acid, which could be prepared by adding 60 grams of acetic acid (1 mole, or 6.02×10^{23} molecules) to 1 L of water (which corresponds to 55.5 moles; see p. 152). Once mixed, there is an average of about 55 water molecules for every acetic acid molecule, and the molecules are all

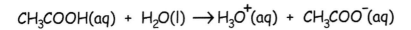

$$CH_3COOH(aq) + H_2O(l) \rightarrow H_3O^+(aq) + CH_3COO^-(aq)$$

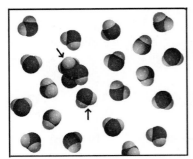 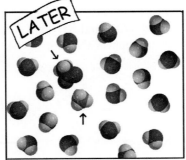

Acetic acid donates a proton to water
to produce hydronium ion and acetate ion.

constantly moving and bumping into each other. In order for the reaction to occur, an acetic acid molecule must bump into a water molecule with the right orientation. That is, the acidic proton must bump into a lone pair on a water molecule.

Although the proper orientation is necessary, it doesn't guarantee that a reaction will occur because the molecules must also possess the right amount of energy. The reverse reaction can occur when a hydrogen from hydronium ion bumps into the oxygen on the acetate ion. Initially the rate of the forward reaction is faster because the concentrations of acetic acid and water are high and there is no acetate around to react in the reverse direction. As time passes, the concentrations of both acetate and hydronium ions increase due to the forward reaction, and thus the rate of the reverse reaction increases. At the same time, the forward reaction rate will decrease because the concentrations of reactants are decreasing (although the decrease in water's concentration is negligible) until the reverse rate is equal to the forward rate and the reaction is at chemical equilibrium.

The concentration of acetic acid in our example is about the same as it is in vinegar, which is simply a 1 M solution of acetic acid in water. In every bottle of vinegar, acetic acid is at chemical equilibrium with its conjugate base acetate ion, and the two opposing reactions we previously described are all happening at the same rate. The next time you see a bottle of vinegar, imagine the dynamic molecular process we have just described occurring endlessly inside the bottle.

An example of a reversible reaction that occurs in the gas phase is the formation of dinitrogen tetroxide (N_2O_4) from nitrogen dioxide (NO_2):

$$2NO_2(g) \rightleftharpoons N_2O_4(g)$$

Unlike the reaction of colorless acetic acid with colorless water, we can actually see the progress of this reaction because N_2O_4 is colorless and NO_2 has a dark brown color (which makes it sometimes visible in polluted air). If we inject NO_2 into an evacuated flask, the dark brown color gets lighter as the reactant is converted to colorless product. The intensity of the brown color decreases because we are replacing the colored molecules with colorless ones. At the molecular level, the reaction occurs when two NO_2 molecules, each with an unpaired electron, collide such that the two electrons form a covalent bond (Figure 8.1). The reverse reaction occurs when N_2O_4 collides with another molecule with enough energy to break the bond between the two nitrogens. Eventually the rate of the reverse reaction catches up to the rate of the forward reaction, and the system reaches equilibrium. At this point, there is no net change in the concentrations of the two species, and the color stays the same—a lighter shade of brown.

What do you think would happen if we injected N_2O_4 into an evacuated flask? Some brown color appears immediately, indicating the formation of NO_2 molecules. The color intensifies as the dissociation of N_2O_4 continues until equilibrium is reached. Beyond that point, no further change in color is evident because the concentrations of both N_2O_4 and NO_2 remain constant. The reaction is clearly reversible because a pure component (N_2O_4 or NO_2) reacts to give the other gas. Another way

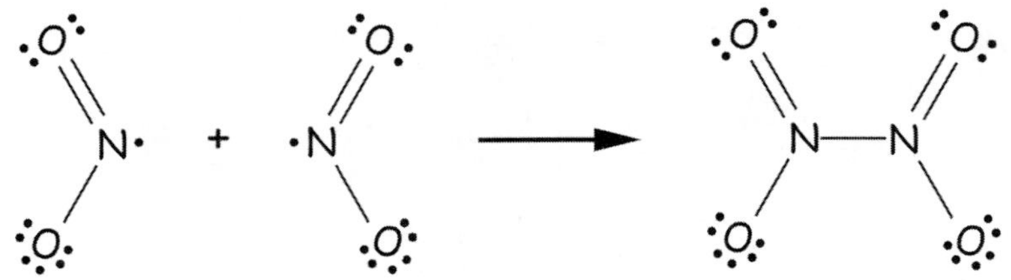

Figure 8.1   Formation of N_2O_4.

to create an equilibrium state is to start with a mixture of NO_2 and N_2O_4 and monitor the system until the color stops changing. Again, it's important to keep in mind that at equilibrium, the conversion of N_2O_4 to NO_2 and that of NO_2 to N_2O_4 is still going on. The reason we do not see a color change is that the two rates are equal— the removal of NO_2 molecules takes place as fast as the production of NO_2 molecules, and N_2O_4 molecules are formed as quickly as they dissociate.

The Equilibrium Constant | *The big K*

Depending on the initial conditions, the concentrations of NO_2 and N_2O_4 can vary greatly at equilibrium. Table 8.1 shows experimental data for the NO_2-N_2O_4 system at 25°C. The gas concentrations are given in molarity, which can be calculated from the number of moles of the gases present initially and at equilibrium and the volume of the flask in liters. Inspection of the data reveals no obvious relationship between the amounts of the two reacting species, and the ratio of their concentrations is highly variable. However, regardless of the initial concentrations of the gases, at equilibrium the ratio of the N_2O_4 concentration over the NO_2 concentration squared gives a nearly constant value that averages 216:

$$\frac{[N_2O_4]}{[NO_2]^2} = 216 = K$$

We call this quotient that has a constant value at equilibrium the **equilibrium constant,** and it is denoted with a capital **K.** Although both $[NO_2]$ and $[N_2O_4]$ have units of mol/L, by convention, K is treated as a dimensionless quantity. We will adhere to this practice for all cases involving chemical equilibrium.

While it's not obvious why the concentration of NO_2 must be squared to give this constant relationship, note that the superscript 2 comes from the stoichiometric coefficient for NO_2 in the balanced equation. It turns out that for the countless reversible reactions that have been studied, a constant value is obtained when the product of the equilibrium concentrations of the products, each raised to a power equal to its coefficient in the balanced equation, is divided by the product of the equilibrium concentrations of reactants, each raised to a power equal to its coefficient. Although there can be any number of reactants and products in a reaction, the relationship between a chemical equation and the equilibrium constant can be effectively described by considering the generalized equation:

$$a\text{A} + b\text{B} \rightleftharpoons c\text{C} + d\text{D}$$

Table 8.1 Experimental Data for the NO_2-N_2O_4 System at 25°C

INITIAL CONCENTRATIONS (*M*)		EQUILIBRIUM CONCENTRATIONS (*M*)		RATIO OF CONCENTRATIONS AT EQUILIBRIUM	
$[NO_2]$	$[N_2O_4]$	$[NO_2]$	$[N_2O_4]$	$\dfrac{[N_2O_4]}{[NO_2]}$	$\dfrac{[N_2O_4]}{[NO_2]^2}$
0.000	0.670	0.0547	0.643	11.7	215
0.0500	0.446	0.0457	0.448	9.80	215
0.0300	0.500	0.0475	0.491	10.3	217
0.0400	0.600	0.0523	0.594	11.4	217
0.200	0.000	0.0204	0.0898	4.41	216

where A and B are reactants, C and D are products, and *a, b, c,* and *d* are the coefficients in the balanced equation. By definition, the equilibrium constant expression is given by

$$K = \frac{[C]^c [D]^d}{[A]^a [B]^b}$$

This equation is the mathematical expression of the **law of mass action,** formulated by the Norwegian chemists ***Cato Guldberg*** and ***Peter Waage*** in 1864.

Writing the equilibrium constant expressions for chemical reactions is straightforward, although there is one important rule to know. The concentrations of pure liquids and pure solids are not included in the equilibrium constant expression. Consider the situation in which we heat solid ammonium chloride (NH_4Cl) in a closed container. The compound decomposes to produce two gases, ammonia (NH_3) and hydrogen chloride (HCl). At a given temperature, the following equilibrium is established:

$$NH_4Cl(s) \rightleftharpoons NH_3(g) + HCl(g)$$

The equilibrium constant (K') may be expressed as

$$K' = \frac{[NH_3][HCl]}{[NH_4Cl]}$$

(You will see why we have called this quotient K'.) Note that the $[NH_4Cl]$ term represents the concentration of a solid in moles per liter, which may seem a bit strange to you. It turns out that a solid's concentration, like its density (measured in g/cm^3), is itself a constant. This is so because no matter how much or how little of the solid is present, the ratio

$$\frac{\text{moles of } NH_4Cl}{\text{volume of } NH_4Cl}$$

is always the same. Because the product of two constants, K' and $[NH_4Cl]$, is also a constant, it is convenient to rewrite the equilibrium constant as

$$K'[NH_4Cl] = K = [NH_3][HCl]$$

where the concentration of the solid is omitted from the expression. As you will see, **the applications of the equilibrium constant deal only with the concentrations of species whose concentration can vary (that is, gases and aqueous solutions).**

Most of the chemical equilibria you will encounter in introductory chemistry involve water, either in acid-base reactions or precipitation reactions. It's possible to calculate the molar concentration of water using its density. Assuming a density of 1 g/mL, 1 liter of water weighs 1000 g. From the molar mass of water (18.02 g/mol), the number of moles of water is 1000 g/(18.02 g/mol) or 55.5 mol. Therefore, the concentration in moles per liter is 55.5 mol/L, or 55.5 *M*.

Consider the ionization of acetic acid in water:

$$CH_3COOH(aq) + H_2O(l) \rightleftharpoons CH_3COO^-(aq) + H_3O^+(aq)$$

According to the law of mass action, we write the expression for K' as

$$K' = \frac{[H_3O^+][CH_3COO^-]}{[CH_3COOH][H_2O]}$$

Because the concentration of water is so much greater than all of the other species, it is essentially constant at 55.5 *M*. This is generally true for all acid-base reactions,

Table 8.2 Some Equilibrium Constant Expressions

Reaction	K Expression
$2CO_2(g) \rightleftharpoons 2CO(g) + O_2(g)$	$K = \dfrac{[CO]^2[O_2]}{[CO_2]^2}$
$N_2(g) + 3H_2(g) \rightleftharpoons 2NH_3(g)$	$K = \dfrac{[NH_3]^2}{[N_2][H_2]^3}$
$HF(aq) + H_2O(l) \rightleftharpoons H_3O^+(aq) + F^-(aq)$	$K = \dfrac{[H_3O^+][F^-]}{[HF]}$
$NH_3(aq) + H_2O(l) \rightleftharpoons NH_4^+(aq) + OH^-(aq)$	$K = \dfrac{[NH_4^+][OH^-]}{[NH_3]}$
$Ag^+(aq) + Cl^-(aq) \rightleftharpoons AgCl(s)$	$K = \dfrac{1}{[Ag^+][Cl^-]}$
$CaCO(s) \rightleftharpoons CaO(s) + CO_2(g)$	$K = [CO_2]$

so we treat the concentration of water as a constant as we did for solid ammonium chloride and omit it in the equilibrium constant expression:

$$K = K'[H_2O] = \frac{[H_3O^+][CH_3COO^-]}{[CH_3COOH]}$$

Table 8.2 gives some examples of the equilibrium constant expressions for a variety of chemical reactions.

You may wonder why the concentrations of reacting species are raised to a power equal to their coefficients in the balanced equation. As it turns out, this rule comes from the dependence of reaction rate on concentration, a concept you should be familiar with by now. Let's first consider a general reaction where the coefficients in the balanced equation are all one. Suppose in a certain reaction, molecule A collides and reacts with molecule B to give molecules C and D:

$$A + B \rightleftharpoons C + D$$

The forward rate is proportional to the concentrations of A and B, and the reverse rate is proportional to the concentrations of C and D:

$$\text{rate}_f \propto [A][B] \quad \text{and} \quad \text{rate}_r \propto [C][D]$$

We can substitute proportionality constants k_f and k_r, which we will call rate constants, to get the equations

$$\text{rate}_f = k_f[A][B] \quad \text{and} \quad \text{rate}_r = k_r[C][D]$$

At equilibrium, we know the rates of the two opposing reactions are equal:

$$k_f[A][B] = k_r[C][D]$$

Rearranging to get the concentrations of products in the numerator and the concentrations of reactants in the numerator,

$$\frac{k_f}{k_r} = \frac{[C][D]}{[A][B]} = K$$

Thus, the ratio of the forward and reverse rate constants is the same as the equilibrium constant, K.

Now let's look at a general reaction in which the coefficients are not all one:

$$A + A \rightleftharpoons C + D \qquad \text{or} \qquad 2A \rightleftharpoons C + D$$

In this reaction, the reverse rate is the same as in the previous example, but the forward rate is given by

$$\text{rate}_f = k_f[A][A] \qquad \text{or} \qquad \text{rate}_f = k_f[A]^2$$

Because two molecules of A must react, the rate depends on the concentration of A raised to the power of 2, the coefficient in the balanced equation. Treating the forward and reverse rates as we did previously for the equilibrium condition, we see that the equilibrium constant shows the same dependence on the concentration of A:

$$\frac{k_f}{k_r} = \frac{[C][D]}{[A]^2} = K$$

Knowing the relationship between the equilibrium constant and the rates of the opposing processes should give you a better understanding of the equilibrium constant, but this knowledge is not needed to write the expression for K for chemical reactions or to solve equilibrium problems.

Here is some additional information you should know about writing equilibrium constant expressions. First, the terms *reactants* and *products* for a reversible reaction are arbitrary. It depends on how the equation is written. Therefore, NO_2 is viewed as reactant if the equation is

$$2NO_2(g) \rightleftharpoons N_2O_4(g) \qquad K_1 = \frac{[N_2O_4]}{[NO_2]^2}$$

and product if we reverse the equation

$$N_2O_4(g) \rightleftharpoons 2NO_2(g) \qquad K_2 = \frac{[NO_2]^2}{[N_2O_4]}$$

As you can see, the two equilibrium constants are related by the equation $K_1 K_2 = 1$ or $K_1 = 1/K_2$.

Determining *K*

How does one determine the value of the equilibrium constant for a particular reaction? As you might expect, you must have a way to measure the amount of at least one of the reacting species. There are countless methods for doing this, depending on the nature of reacting species involved. For reactions in which either the reactants or products are colored (that is, they absorb visible light of a certain wavelength), the amount of light absorbed can be quantified using an instrument called a *spectrophotometer*. The amount of light absorbed by a substance is directly proportional to its concentration (more molecules absorb more photons). In the $N_2O_4 \rightleftharpoons 2NO_2$ reaction, a spectrophotometer can be used to measure the concentration of NO_2 over time. As the reaction proceeds, the absorbance reading on the spectrophotometer would continue to increase (if initially only N_2O_4 was present) over time until equilibrium is reached, at which point there is no further change in the concentration and the absorbance remains constant. The absorbance at this point will give the equilibrium concentration of NO_2. Using simple algebra, the equilibrium concentration of N_2O_4 can be calculated if you know its initial concentration. The equilibrium constant is then readily determined by plugging the equilibrium concentrations into the expression for K, as shown in the following example.

EXAMPLE 8.1

Calculate the equilibrium constant for the $N_2O_4 \rightleftharpoons 2NO_2$ reaction at 25°C if the initial concentration of N_2O_4 in a 1-liter flask was 0.600 M and the absorbance of the light brown mixture at equilibrium indicated a NO_2 concentration of 0.0516 M.

Answer   To solve this problem, we calculate the equilibrium concentration of N_2O_4 and then plug the equilibrium concentrations of both species into the expression for K. According to the balanced equation, 2 moles of NO_2 are produced for each mole of N_2O_4 that reacts. Thus, the amount of N_2O_4 that reacted was one-half of the 0.0516 mole of NO_2 produced, or 0.0258 mole of N_2O_4. The equilibrium concentration of N_2O_4 is then 0.600 − 0.0258, or 0.574 M. Solving for K, we get

$$K = \frac{[NO_2]^2}{[N_2O_4]} = \frac{(0.0516)^2}{0.574} = 4.63 \times 10^{-3}$$

Go to the website to practice solving more problems.

Applications of the Equilibrium Constant

If the equilibrium constant has been determined for a particular reaction, chemists studying the reaction can use it to predict

- The net direction the reaction will move toward equilibrium if the initial concentrations of reacting species are known.
- The concentrations of the reacting species at equilibrium.
- The effects of changing concentrations, pressure, or temperature on a system already at equilibrium.

Predicting the direction of reaction not at equilibrium

Which way will it go—left to right or right to left?

We can easily determine if a reaction is at equilibrium if we know the concentrations of reacting species by plugging the concentrations into the equilibrium constant expression. If the quotient is not numerically equal to the value of K, then it's not at equilibrium, and therefore the reaction will proceed from left to right or in the reverse direction until the quotient equals K. Although mathematically the quotient has the same form as that for the equilibrium constant, we can't call the quotient K because the reaction is not at equilibrium. Therefore, we use the letter Q, for **reaction quotient.** By comparing the numerical values of K and Q, we can predict which way the reaction will go.

Suppose we have a flask containing two gases at the following initial concentrations: $[NO_2]_0 = 0.018\ M$ and $[N_2O_4]_0 = 0.16\ M$ at 25°C. The subscript 0 denotes the initial concentrations. To see if the reaction $[N_2O_4 \rightleftharpoons 2NO_2]$ has reached equilibrium, we can plug these concentrations into the reaction quotient and compare the value with K.

$$Q = \frac{[NO_2]_0^2}{[N_2O_4]_0}$$

$$= \frac{(0.018)^2}{0.16} = 2.0 \times 10^{-3}$$

Because Q is smaller than the value for K (4.63×10^{-3}), we conclude that the reaction is not at equilibrium. To reach equilibrium, the reaction will go from left to right to produce more NO_2.

Remember that the only difference between Q and K is that the former is used when the concentrations are not at equilibrium. In comparing Q with K, there are three possibilities:

1. **$Q > K$. The product:reactant ratio is greater than it is at equilibrium, so the reaction must produce more reactants to reach equilibrium.** In the preceding reaction, there is too much NO_2 or not enough N_2O_4 in the reacting mixture. To reach equilibrium, the net reaction will be from right to left, depleting NO_2 and forming N_2O_4 until equilibrium is reached. At that point, the forward and reverse rates will be equal, and there will be no net reaction in either direction.
2. **$Q = K$. The reaction is at equilibrium.** In this case, nothing happens.
3. **$Q < K$. The product: reactant ratio is smaller than it is at equilibrium, so the reaction must produce more products to reach equilibrium.** This is the opposite of Case 1. For the preceding reaction, this would mean we have too much N_2O_4 (or not enough NO_2), so the net reaction will be from left to right, depleting N_2O_4 and forming NO_2 until equilibrium is reached.

Calculating Equilibrium Concentrations

How much reactant and product will there be at equilibrium?

Knowing the equilibrium constant enables us to calculate the concentrations of the reacting species at equilibrium if we know the initial concentrations. Let's again use the previous example, that is, the initial concentrations of $[NO_2]_0 = 0.018\ M$, and $[N_2O_4]_0 = 0.16\ M$. From the reaction quotient, we already saw that to reach equilibrium, the concentration of NO_2 will increase and that of N_2O_4 will decrease. If we assume the concentration of N_2O_4 is decreased by $x\ M$ at equilibrium, then it follows that the concentration of NO_2 must increase by $2x\ M$ (because for every N_2O_4 molecule decomposed, two molecules of NO_2 are formed). The coefficient for x will always be the same as the corresponding coefficient in the balanced equation. There is an easy way to keep track of how concentrations change to reach equilibrium. We set up a table to show the initial conditions, changes, and the final (equilibrium) conditions:

	$N_2O_4(g) \rightleftharpoons$	$2NO_2(g)$
Initial (M):	0.16	0.018
Change (M):	$-x$	$+2x$
Equilibrium (M):	$0.16 - x$	$0.018 + 2x$

A minus sign denotes a decrease and a plus sign denotes an increase in the concentration. So if we can solve for x, we will know the equilibrium concentrations of NO_2 and N_2O_4. This we can do because we know the value of K, so we write

$$K = \frac{[NO_2]^2}{[N_2O_4]}$$

$$4.63 \times 10^{-3} = \frac{(0.018 + 2x)^2}{(0.16 - x)}$$

or $\qquad 7.41 \times 10^{-4} - 4.63 \times 10^{-3}x = 3.24 \times 10^{-4} + 0.072x + 4x^2$

$$4x^2 + 0.0766x - 4.17 \times 10^{-4} = 0$$

As you may have noticed, this is a quadratic equation of the form

$$ax^2 + bx + c = 0$$

The formula for solving for x is

$$x = \frac{-b \pm \sqrt{b^2 - 4ac}}{2a}$$

Here we have $a = 4$, $b = 0.0766$, and $c = -4.17 \times 10^{-4}$ so that

$$x = \frac{-0.0766 \pm \sqrt{(0.0766)^2 - 4(4)(-4.17 \times 10^{-4})}}{2(4)}$$

and

$$x = -0.024\ M \quad \text{or} \quad x = 0.0044\ M$$

The first answer is physically impossible because it predicts the equilibrium concentration of N_2O_4 will be greater than the initial concentration. From the second answer for x, we calculate the equilibrium concentrations as follows:

$$[NO_2] = (0.018 + 2x)\ M$$
$$= (0.018 + 2 \times 0.0044)\ M = 0.027\ M$$
$$[N_2O_4] = (0.16 - x)\ M$$
$$= (0.16 - 0.0044)\ M \approx 0.16\ M$$

where the $\approx$ sign means "approximately equal to." This method is sometimes referred to as the ICE method, where the acronym stands for Initial, Change, and Equilibrium. You can use the ICE method to solve all equilibrium problems you encounter.

Go to the website to practice solving more problems.

Le Chatelier's Principle

What happens when we disturb a system at equilibrium?

Chemists often want to know what happens to an equilibrium mixture if it is disturbed in some way. The disturbance can be the addition or removal of some reacting species, heating or cooling, or a change in pressure and volume. As you will see, knowing how an equilibrium state responds to such disturbances can help chemists determine the conditions that will optimize the yield of a desired compound.

IF A STRESS IS PLACED ON A REACTION AT EQUILIBRIUM, THE REACTION WILL PROCEED IN THE DIRECTION THAT WILL RELIEVE THE STRESS.

Le Chatelier describes his principle.

The French chemist **Henri Louis Le Chatelier** proposed a general rule that helps us to predict the direction toward which an equilibrium reaction will move when a change in concentration, pressure, volume, or temperature occurs. This rule, known as **Le Chatelier's principle,** states that **if an external stress is applied to a system at equilibrium, the system adjusts in a way to relieve the stress as the system reaches a new equilibrium position.** The word "stress" here means a change in concentration, pressure, volume, or temperature that removes the system from the equilibrium state. To illustrate the principle, let's continue to focus on the same reaction:

$$N_2O_4(g) \rightleftharpoons 2NO_2(g)$$

Changes in Concentration.   Suppose initially we have a mixture of NO_2 and N_2O_4 gases at equilibrium in a flask. Now we inject some N_2O_4 into the mixture. (What happens to the color?) Instantly, the equilibrium condition is destroyed because the concentration of N_2O_4 is increased. Knowing the amount of N_2O_4 added, we can calculate the reaction quotient Q to show that it is smaller than K

$$Q = \frac{[NO_2]_0}{[N_2O_4]_0} < K$$

because the denominator is larger than the equilibrium concentration. The subscript 0 refers to the concentrations right after some N_2O_4 has been added to the flask. To offset this stress—that is, the addition of extra N_2O_4—some N_2O_4 will decompose to form NO_2 until the concentrations of the gases are such that the following ratio is equal to K:

$$\frac{[NO_2]^2}{[N_2O_4]} = K$$

So an addition of N_2O_4 will shift the equilbrium to the right, forming more NO_2 molecules.

Eventually, both the concentrations of NO_2 and N_2O_4 will be higher, but the ratio $[NO_2]^2/[N_2O_4]$ is such that it is again equal to K at equilibrium. Conversely, an addition of NO_2 will shift the equilibrium to the left, forming more N_2O_4 molecules. By the same token, if we remove some N_2O_4 from the equilibrium mixture, more NO_2 molecules will combine to form N_2O_4 as a way to offset the stress, and the net reaction will shift to the left. What would happen if we remove some of the NO_2 molecules from the equilibrium mixture? In each case, you can predict the direction of the net reaction by comparing Q with K.

Changes in Pressure and Volume.   Consider the following set up in which a mixture of NO_2 and N_2O_4 gases at equilibrium is placed in a cylinder fitted with a movable piston. The upward or downward movement of the piston will increase or decrease the volume of the cylinder. Suppose the piston is pushed downward and the volume decreases. Because the numbers of moles of the gases remain the same at that instant but the volume has shrunk, the concentrations (in moles per liter) of both gases will go up.

How will the system adjust to this stress? According to Le Chatelier, the system should respond to an increase in pressure (the stress) by moving in a direction that will reduce pressure (stress relief). The only way that a reaction can cause a decrease in pressure is if it proceeds to the side with fewer moles of gas. From the balanced equation, we see that it takes two molecules of NO_2 to produce one molecule of N_2O_4. Therefore, we have a net reaction from right to left. This will produce fewer gas molecules and, therefore, less pressure. Only reactions involving unequal numbers of

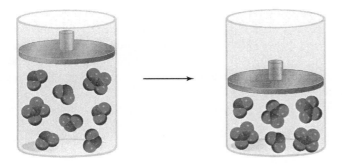

moles of gases on the left and right sides of the equation are affected by changes in pressure and volume. Changes in pressure and volume will have no effect on a reaction that produces equal numbers of molecules in either direction (left to right and right to left); for example,

$$H_2(g) + I_2(g) \rightleftharpoons 2HI(g)$$

In addition, reactions in condensed phases (liquids and solids) are unaffected by pressure because solids and liquids are highly incompressible, so their volumes remain the same.

As with the stress of changing concentration, the response to the pressure and volume change is to restore K. You can understand why this is the case if you consider what happens to Q when the pressure and volume are changed. For example, doubling the pressure (by halving the volume) would double the concentrations of both gases in the NO_2-N_2O_4 system. Because the doubled concentration of NO_2 is squared in the numerator and the doubled N_2O_4 concentration is not, $Q > K$. The reaction must proceed to the right, that is, $2NO_2 \rightarrow N_2O_4$ until $Q = K$ and a new equilibrium state is reached. Conversely, raising the piston will increase the volume and decrease the concentrations of the gases. In that case, the equilibrium will shift from left to right, that is, $N_2O_4 \rightarrow 2NO_2$ to produce more gas molecules (and greater pressure) until a new equilibrium state is reached. Try doubling or halving some of the equilibrium concentrations from Table 8.1 to prove to yourself that the value of K is destroyed.

Changes in Temperature. This time we have the NO_2-N_2O_4 equilibrium mixture in a glass flask of constant volume. What happens when the mixture is heated? According to Le Chatelier, the system should respond to the stress (an increase in temperature) by absorbing the heat. You know from Chapter 5 that for reversible reactions, heat is released in one direction and an equal amount of heat is absorbed in the opposite direction. We know that the decomposition of N_2O_4 to form NO_2 molecules requires the breaking of a chemical bond, so energy must be supplied from an outside source. Thus, the following reaction is endothermic:

$$\text{heat} + N_2O_4(g) \longrightarrow 2NO_2(g) \qquad \Delta H = +58.0 \text{ kJ}$$

Conversely, the formation of N_2O_4 when two NO_2 molecules combine is a bond-forming and hence exothermic process:

$$2NO_2(g) \longrightarrow N_2O_4(g) + \text{heat} \qquad \Delta H = -58.0 \text{ kJ}$$

As a way to relieve the stress (heating), the net reaction will be from left to right (the endothermic direction) to absorb the added heat until a new equilibrium is reached at a higher temperature. The important thing to note is that, unlike Cases 1 and 2, not only will we have a new equilibrium position in this case (meaning altered concentrations for NO_2 and N_2O_4), but the value of the equilibrium constant itself is changed (increased in this case. Why?). If the equilibrium mixture is cooled, then we predict the stress (cooling) will favor the exothermic reaction and hence the formation of N_2O_4. Again, we will have a different (decreased) equilibrium constant at a lower temperature.

(*Check the website for problems on factors affecting the equilibrium position.*)

Acid-Base Equilibria

One of the most important topics in chemistry is *acid-base equilibria*. Knowing the fundamentals of acid-base equilibria enables us to understand natural phenomena such as the effects of acid rain on both living and nonliving matter or the chemistry of the molecules that create and maintain life. An acid-base equilibrium is a bit more complicated than the gas-phase equilibrium we just studied because it always involves more than one reversible reaction occurring simultaneously. In addition to the equilibria that directly involves the acids or bases themselves, there is also the equilibrium involving water. Before you read any further, it might be helpful to reread the section in Chapter 3 on acid-base reactions (pp. 35–40).

The Water Equilibrium

Water ionizes with itself.

Water is involved in all aqueous acid-base equilibria because it can act as either a weak acid or a weak base. If an acid is added to water, water molecules can accept protons donated by the acid, that is, they behave as a Brønsted base. Once water accepts a proton, it becomes a hydronium ion (H_3O^+), as shown in the generalized equation

$$HB(aq) + H_2O(l) \rightleftharpoons H_3O^+(aq) + B^-(aq)$$
$$\text{acid}_1 \qquad \text{base}_2 \qquad \text{acid}_2 \qquad \text{base}_1$$

In this equation, HB represents any acid, and B^- represents the base that is produced after the acid has donated its proton. We call the base in this case the **conjugate base** of the acid HB (although it's just a base like any other base). The subscripts 1 and 2 help us keep track of an acid and its conjugate base.

If a base is added to water, water molecules will donate a proton to the base, that is, they behave as a Brønsted acid, and become the hydroxide ion (OH^-), as shown in the generalized equation

$$B^-(aq) + H_2O(l) \rightleftharpoons OH^-(aq) + HB(aq)$$
$$\text{base}_1 \qquad \text{acid}_2 \qquad \text{base}_2 \qquad \text{acid}_1$$

where B^- represents any base, which usually has a negative charge, and HB is the acid formed after the base accepts a proton from water. We call the acid thus formed the **conjugate acid** of the base B^- (although it's just an acid like any other acid).

Because water can either donate or accept a proton, a certain amount of proton transfer occurs even in pure water where one water molecule bumps into another with enough energy and the right orientations and donates a proton:

$$H_2O(l) + H_2O(l) \rightleftharpoons H_3O^+(aq) + OH^-(aq)$$

We will refer to this reversible reaction as the *water equilibrium,* for which the equilibrium constant is given by

$$K = [H_3O^+][OH^-]$$

Remember that because the concentration of water (55.5 *M*) is very large, we treat it as a constant and don't include it in the equilibrium constant expression. The water equilibrium is so important that it has been given its own special *K,* which we denote as K_w. At 25°C,

$$K_w = [H_3O^+][OH^-] = 1.0 \times 10^{-14}$$

(As stated on p. 151, by convention the equilibrium constants for acid-base reactions have no units.) From the stoichiometry, we see that $[H_3O^+] = [OH^-]$, so in water we have

$$[H_3O^+] = 1.0 \times 10^{-7} \, M$$

$$[OH^-] = 1.0 \times 10^{-7} \, M$$

These are very low concentrations; in fact, only one in 10 million water molecules undergoes ionization. For this reason, pure water is an extremely poor electrical conductor because it contains very few ions.

One of the most important features of acid-base equilibria is that the addition of acid or base to water affects the water equilibrium. For example, if an acid is added, the hydronium ion concentration increases, resulting in a situation where $Q_w > K_w$. According to Le Chatelier's principle, the increase in $[H_3O^+]$ will cause the water equilibrium to move to the left to consume the added H_3O^+ ions, thereby relieving the stress. The shift towards reactants must also reduce the hydroxide ion concentration until a new equilibrium is reached when we again have

$$[H_3O^+][OH^-] = 1.0 \times 10^{-14}$$

But here we have $[H_3O^+] > [OH^-]$. We can change the concentration of either H_3O^+ or OH^- ions in solution (by adding acid or base), but we cannot vary both of them independently. If we add enough acid to the solution so that $[H_3O^+] = 1.0 \times 10^{-6} \, M$, say, then the OH^- concentration must change to

$$[OH^-] = \frac{K_W}{[H_3O^+]} = \frac{1.0 \times 10^{-14}}{1.0 \times 10^{-6}} = 1.0 \times 10^{-8} \, M$$

What happens to the hydronium ion concentration when a base is added to water?

Whenever $[H_3O^+] = [OH^-]$, as in pure water, the aqueous solution is said to be neutral. As you have seen, an acidic solution contains an excess of H_3O^+ ions, and $[H_3O^+] > [OH^-]$. In a basic solution, there is an excess of hydroxide ions, so $[H_3O^+] < [OH^-]$.

pH | *the negative log of* [*H₃O⁺*]

Because $[H_3O^+]$ is usually very small in water or even in acid solutions, the Danish biochemist **Soren Sorensen** introduced a quantity called **pH,** which he called the "hydrogen ion exponent." It is defined as the negative logarithm of the hydronium ion concentration:

$$pH = -\log [H_3O^+]$$

162

In pure water, then, the pH is given by

$$pH = -\log (1.0 \times 10^{-7})$$
$$= 7.00$$

THE IMPORTANT THING TO
REMEMBER IS THAT A LOW pH
VALUE MEANS THE HYDRONIUM
ION CONCENTRATION IS HIGH

Tired of dealing with very small numbers,
Sorensen introduced pH and defined it as the
negative log of the hydronium ion concentration.

Note that in calculating the pH of a solution, we use only the numerical part of $[H_3O^+]$ because we cannot take the logarithm of units. Depending on the relative amounts of H_3O^+ and OH^- present, a solution can be labeled as acidic, basic, or neutral according to the following:

Acidic solution: $[H_3O^+] > 1.0 \times 10^{-7}\ M$ pH < 7.00

Basic solution: $[H_3O^+] < 1.0 \times 10^{-7}\ M$ pH > 7.00

Neutral solution: $[H_3O^+] = 1.0 \times 10^{-7}\ M$ pH $= 7.00$

pH values usually range from 0 to 15, although it is possible to have solutions with a negative pH (like a highly concentrated solution of a strong acid). Remember that pH decreases (becoming more acidic) as $[H_3O^+]$ increases. So a low pH means the solution is very acidic.

8.2

EXAMPLE

The concentration of H_3O^+ ions in a bottle of wine that has been exposed to air for several weeks is 2.5×10^{-3} M. Calculate the pH of the solution.

Answer   *The pH of the wine is*

$$pH = -log\,[H_3O^+]$$
$$= -log\,(2.5 \times 10^{-3})$$
$$= 2.60$$

This "acidic" wine is due to the conversion of ethanol to acetic acid.

The pH of rainwater collected in a certain region of Massachusetts on a particular day was 4.82. Calculate the H_3O^+ ion concentration of the rainwater.

Answer Here we are given the pH of a solution and asked to calculate $[H_3O^+]$. Because pH is defined as $pH = -\log [H_3O^+]$, we can solve for $[H_3O^+]$ by taking the antilog of the pH. We write

$$pH = -\log [H_3O^+] = 4.82$$

Therefore,

$$\log [H_3O^+] = -4.82$$

$$[H_3O^+] = 10^{-4.82} = 1.5 \times 10^{-5} \, M$$

(Check the website for more problems on pH.)

Solving Acid-Base Problems | *It's just equilibria*

You now have the foundation needed for solving acid-base problems (most of which involve calculating the pH of a particular solution). As we mentioned earlier, acid-base problems are generally more involved than the gas-phase equilibria because there is more than one type of equilibrium in an aqueous solution. A key step in learning how to solve these problems is knowing which of the species are important and which can be ignored. A systematic way to do this is to first list the major species in solution and then examine the influence each of them has on the pH of the solution.

Strong Acids and Strong Bases

As far as calculating the pH of a solution is concerned, strong acids and strong bases are easier to deal with than weak acids and weak bases because we assume they ionize completely. (Strong acid ionizations are actually reversible to a very small extent, but it's negligible.) Suppose we wish to calculate the pH of a 0.20 *M* HCl solution. How should we go about it? The following steps provide a systematic approach to solving the problem.

Step 1: List the species present initially, before ionization has begun. Here we have HCl and H_2O.

Step 2: Write the reactions that occur in solution. There are two:

$$HCl(aq) + H_2O(l) \longrightarrow H_3O^+(aq) + Cl^-(aq) \quad K = \text{very large}$$

$$H_2O(l) + H_2O(l) \rightleftharpoons H_3O^+(aq) + OH^-(aq) \quad K_w = 1.0 \times 10^{-14}$$

Note that because HCl is a strong acid, its ionization is complete, and we use the single arrow.

Step 3: What are the species in solution at equilibrium? The species are H_3O^+, Cl^-, OH^-, and H_2O. Note that no HCl is left at this point. Both of the preceding reactions contribute to H_3O^+ ion concentration, but because the amount of hydronium ion produced by the ionization of water is negligible compared with that produced by HCl, we can ignore it. The Cl^- ion is the conjugate base of a strong acid and therefore has no tendency to take a proton from any acid. It certainly will not react with the weak acid water, and therefore it is present only as a spectator ion; that is, it does not affect the pH of the solution. The OH^- ion is present in very small quantity and can also be ignored for now.

Step 4: Solve for $[H_3O^+]$ by using the ICE method shown earlier. Here we have

	$HCl(aq) + H_2O(l) \longrightarrow H_3O^+(aq) + Cl^-(aq)$		
Initial (M):	0.20	0	0
Change (M):	-0.20	0.20	0.20
Equilibrium (M):	0	0.20	0.20

The pH of the solution is given by

$$pH = -\log (0.20)$$
$$= 0.70$$

As you can see, this systematic approach wasn't really necessary for HCl because the hydronium ion concentration at equilibrium is equal to the initial concentration of the acid. In fact, for all the strong monoprotic acids (there are five, remember?), the pH of the solution is therefore the negative logarithm of the initial acid concentration.

To calculate the pH of basic solution, you must first determine the hydroxide ion concentration and then calculate the hydronium ion concentration using K_w as previously shown. As we mentioned in Chapter 3, there is only one strong base that you will encounter, the hydroxide ion. Let's say we are interested in the pH of a 3.6 M NaOH solution. Remember that the negatively charged hydroxide ion must be accompanied by a cation. NaOH is very soluble in water, and an aqueous solution of NaOH contains only aqueous Na^+ and OH^- ions. The sodium ion will not react with water to accept or donate a proton—it has no hydrogen to donate, and if it accepted a proton from water, it would become NaH^{2+}, which is highly improbable because like charges repel each other. So like the chloride ion, the sodium ion is just a spectator here. In this case, the only equilibrium is that of water:

$$H_2O(l) + H_2O(l) \rightleftharpoons H_3O^+(aq) + OH^-(aq)$$

According to Le Chatelier's principle, the added hydroxide ion causes a shift to the left until

$$[H_3O^+][OH^-] = 1.0 \times 10^{-14}$$

Because the initial concentration of hydroxide is very large compared to that provided by the ionization of water, we can assume that its equilibrium concentration is the same as the initial concentration of 3.6 M. From the water equilibrium, we write

$$[H_3O^+] = \frac{K_W}{[OH^-]}$$
$$= \frac{1.0 \times 10^{-14}}{3.6}$$
$$= 2.8 \times 10^{-15} \ M$$

so the pH is

$$pH = -\log (2.8 \times 10^{-15})$$
$$= 14.55$$

Weak Acids and Weak Bases

As mentioned earlier, the ionizations of weak acids and weak bases are more involved because their ionizations are not complete. Let's first tackle the problem of calculating the pH of a 0.10 M CH_3COOH (acetic acid) solution. Acetic acid ionizes in solution as follows:

$$CH_3COOH(aq) + H_2O(l) \rightleftharpoons H_3O^+(aq) + CH_3COO^-(aq)$$

As previously shown, the equilibrium constant for this reaction, assuming the water concentration is essentially unaffected, is given by

$$K = \frac{[H_3O^+][CH_3COO^-]}{[CH_3COOH]}$$

Because all acids react with water to produce hydronium ion and the conjugate base, we give the equilibrium constant a special subscript like we did with the water equilibrium:

$$K = K_a = \frac{[H_3O^+][CH_3COOH^-]}{[CH_3COOH]} = 1.8 \times 10^{-5} \text{ at } 25°C$$

where K_a is called the **acid ionization constant.** Remember that K_a is simply the equilibrium constant for the reaction of an acid with water. As such, it's a quantitative measure of acid strength—the larger the K_a value, the stronger the acid. Table 8.3 lists the K_a values for some common acids.

We can now proceed to solve for $[H_3O^+]$ and pH using the following steps.

Step 1: Initially we have CH_3COOH and H_2O present.

Step 2: The reactions that occur in solution are

$$CH_3COOH(aq) + H_2O(l) \rightleftharpoons H_3O^+(aq) + CH_3COO^-(aq) \quad K_a = 1.8 \times 10^{-5}$$

$$H_2O(l) + H_2O(l) \rightleftharpoons H_3O^+(aq) + OH^-(aq) \quad K_w = 1.0 \times 10^{-14}$$

Step 3: At equilibrium, the major species in solution are CH_3COOH, H_3O^+, CH_3COO^-, and H_2O. Note that even though CH_3COOH is a weak acid, it is the main source of H_3O^+ ions, and we can again ignore the ionization of water.

Step 4: We now solve for $[H_3O^+]$ using the ICE method. Let x M be the amount of CH_3COOH that undergoes ionization so that

	$CH_3COOH(aq)$ + $H_2O(l)$ $\rightleftharpoons$	$H_3O^+(aq)$ +	$CH_3COO^-(aq)$
Initial (*M*):	0.10	0	0
Change (*M*):	$-x$	$+x$	$+x$
Equilibrium (*M*):	$0.10 - x$	x	x

The acid ionization constant is given by

$$K_a = \frac{[H_3O^+][CH_3COO^-]}{[CH_3COOH]}$$

$$1.8 \times 10^{-5} = \frac{x^2}{0.10 - x}$$

$$x^2 + 1.8 \times 10^{-5}\, x - 1.8 \times 10^{-6} = 0$$

This is a quadratic equation in x and can be solved as before. There is a short cut, however. Because CH_3COOH ionizes only to a slight extent, x must be small compared to 0.10. Therefore, we can make the approximation

$$0.10 - x \approx 0.10$$

Table 8.3 Ionization Constants of Some Weak Acids and Their Conjugate Bases at 25°C

Name of Acid	Formula	Structure	K_a	Conjugate Base	K_b
Hydrofluoric acid	HF	H—F	7.1×10^{-4}	F^-	1.4×10^{-11}
Nitrous acid	HNO_2	O=N—O—H	4.5×10^{-4}	NO_2	2.2×10^{-11}
Acetylsalicylic acid (aspirin)	$C_9H_8O_4$		3.0×10^{-4}	$C_9H_7O_4^-$	3.3×10^{-11}
Formic acid	HCOOH		1.7×10^{-4}	$HCOO^-$	5.9×10^{-11}
Ascorbic acid*	$C_6H_8O_6$		8.0×10^{-5}	$C_6H_7O_6^-$	1.3×10^{-10}
Benzoic acid	C_6H_5COOH		6.5×10^{-5}	$C_6H_5COO^-$	1.5×10^{-10}
Acetic acid	CH_3COOH		1.8×10^{-5}	CH_3COO^-	5.6×10^{-10}
Hydrocyanic acid	HCN	H—C≡N	4.9×10^{-10}	CN^-	2.0×10^{-5}
Phenol	C_6H_5OH		1.3×10^{-10}	$C_6H_5O^-$	7.7×10^{-5}

*For ascorbic acid it is the upper left OH group that is associated with this ionization constant.

Now the acid ionization constant expression becomes

$$1.8 \times 10^{-5} = \frac{x^2}{0.10}$$

Rearranging, we get

$$x^2 = (0.10)(1.8 \times 10^{-5})$$

$$x = \sqrt{1.8 \times 10^{-6}} = 1.3 \times 10^{-3} \, M$$

At equilibrium, the pH of the solution is

$$pH = -\log(1.3 \times 10^{-3}) = 2.89$$

so the solution is acidic, as we would expect.

How good is the approximation? Because K_a values for weak acids are generally known to an accuracy of only ±5%, it is reasonable to require x to be less than 5% of 0.10, the number from which it is subtracted. In other words, the approximation is valid if the following expression is equal to or less than 5%:

$$\frac{1.3 \times 10^{-3} M}{0.10 \, M} \times 100\% = 1.3\%$$

Thus, the approximation we made is acceptable. In general, you should always take the short cut in solving this kind of problem. But be sure to check the validity of the assumption. If the assumption does not hold, then you must solve for x using the quadratic equation.

Essentially the same approach can be used for weak bases. As another exercise, we will calculate the pH of a 0.20 M ammonia solution. As a weak base, ammonia ionizes in solution as follows:

$$NH_3(aq) + H_2O(l) \rightleftharpoons NH_4^+(aq) + OH^-(aq)$$

and the equilibrium constant for the reaction, again assuming that the water concentration is largely unaffected, is given by

$$K = K_b = \frac{[NH_4^+][OH^-]}{[NH_3]} = 1.8 \times 10^{-5} \quad \text{at 25°C}$$

where K_b is called the **base ionization constant**. (As with K_a, K_b is simply the equilibrium constant for the reaction of a base with water to produce hydroxide ion and its conjugate acid.) K_b is a quantitative measure of base strength—the larger the K_b value, the stronger the base. Table 8.4 lists the K_b values for some common bases. (Note that by coincidence, the K_b for NH_3 is the same as the K_a for CH_3COOH.)

Table 8.4 Ionization Constants of Some Weak Bases and Their Conjugate Acids at 25°C

Name of Base	Formula	Structure	K_b*	Conjugate Acid	K_a
Ethylamine	$C_2H_5NH_2$	$CH_3-CH_2-\overset{..}{N}-H$ with H below	5.6×10^{-4}	$C_2H_5\overset{+}{N}H_3$	1.8×10^{-11}
Methylamine	CH_3NH_2	$CH_3-\overset{..}{N}-H$ with H below	4.4×10^{-4}	$CH_3\overset{+}{N}H_3$	2.3×10^{-11}
Caffeine	$C_8H_{10}N_4O_2$		4.1×10^{-4}	$C_8H_{11}\overset{+}{N}_4O_2$	2.4×10^{-11}
Ammonia	NH_3	$H-\overset{..}{N}-H$ with H below	1.8×10^{-5}	NH_4^+	5.6×10^{-10}
Pyridine	C_5H_5N	ring with N:	1.7×10^{-9}	$C_5H_5\overset{+}{N}H$	5.9×10^{-6}
Aniline	$C_5H_5NH_2$	ring $-\overset{..}{N}-H$ with H below	3.8×10^{-10}	$C_6H_5\overset{+}{N}H_3$	2.6×10^{-5}
Urea	N_2H_4CO	$H-\overset{..}{N}-\overset{O}{\overset{\|}{C}}-\overset{..}{N}-H$	1.5×10^{-14}	$H_2NCO\overset{+}{N}H_3$	0.67

*The nitrogen atom with the lone pair accounts for each compound's basicity.

We now proceed as follows.

Step 1: Initially we have NH_3 and H_2O.

Step 2: The reactions that occur in solution are

$$NH_3(aq) + H_2O(l) \rightleftharpoons NH_4^+(aq) + OH^-(aq)$$

$$H_2O(l) + H_2O(l) \rightleftharpoons H_3O^+(aq) + OH^-(aq)$$

Step 3: At equilibrium, the major species in solution are NH_3, NH_4^+, OH^-, and H_2O. Note that even though NH_3 is a weak base, it is the main source of OH^- ions, and we can again ignore the ionization of water.

Step 4: Let $x\ M$ be the amount of NH_3 ionized. Using the ICE method, we write

$$NH_3(aq) + H_2O(l) \rightleftharpoons NH_4^+(aq) + OH^-(aq)$$

Initial (M):	0.20	0	0
Change (M):	$-x$	$+x$	$+x$
Equilibrium (M):	$0.20 - x$	x	x

The base ionization constant is given by

$$K_b = \frac{[NH_4^+][OH^-]}{[NH_3]}$$

$$1.8 \times 10^{-5} = \frac{x^2}{0.20 - x}$$

We assume that $0.20 - x \approx 0.20$ so that

$$x^2 = (0.20)(1.8 \times 10^{-5}) = 3.6 \times 10^{-6}$$

or

$$x = 1.9 \times 10^{-3}\ M = [OH^-]$$

Now we need to test the approximation by writing

$$\frac{1.9 \times 10^{-3}\ M}{0.20\ M} \times 100\% = 0.95\%$$

so the approximation is valid.

Next we calculate $[H_3O^+]$ using the ion product of water:

$$[H_3O^+] = \frac{K_W}{[OH^-]}$$

$$= \frac{1.0 \times 10^{-14}}{1.9 \times 10^{-3}} = 5.3 \times 10^{-12}\ M$$

The pH of the solution is therefore

$$pH = -\log(5.3 \times 10^{-12}) = 11.28$$

and the solution is basic as we would expect.

As you can see in Table 8.5, most weak bases are anions, so, like hydroxide ion, they must be accompanied by a cation. In fact, most bases exist as the anion components of salts, which you will recall are ionic compounds. When a salt dissolves in water, the pH of the solution may change from 7. This happens when the cation or anion, or

Table 8.5　Relative Strengths of Conjugate Acid-Base Pairs

Acid	Conjugate Base
$HClO_4$ (perchloric acid)	ClO_4^- (perchlorate ion)
HI (hydroiodic acid)	I^- (iodide ion)
HBr (hydrobromic acid)	Br^- (bromide ion)
HCl (hydrochloric acid)	Cl^- (chloride ion)
H_2SO_4 (sulfuric acid)	HSO_4^- (hydrogen sulfate ion)
HNO_3 (nitric acid)	NO_3^- (nitrate ion)
H_3O^+ (hydronium ion)	H_2O (water)
HSO_4^- (hydrogen sulfate ion)	SO_4^{2-} (sulfate ion)
HF (hydrofluoric acid)	F^- (fluoride ion)
HNO_2 (nitrous acid)	NO_2^- (nitrate ion)
$HCOOH$ (formic acid)	$HCOO^-$ (formate ion)
CH_3COOH (acetic acid)	CH_3COO^- (acetate ion)
NH_4^+ (ammonium ion)	NH_3 (ammonia)
HCN (hydrocyanic acid)	CN^- (cyanide ion)
H_2O (water)	OH^- (hydroxide ion)
NH_3 (ammonia)	NH_2^- (amide ion)

Acid strength increases · *Strong acids* · *Weak acids* · *Base strength increases*

both, react with water. This type of acid-base reaction is often called **salt hydrolysis.** But remember that it is simply acid-base equilibria and therefore problems involving salt hydrolysis are solved exactly as we have done previously for weak acids and weak bases. The key is identifying which species will react with water and which will not.

To illustrate salt hydrolysis, consider what happens when sodium acetate (CH_3COONa) is added to water. Being a strong electrolyte, the salt is completely dissociated in solution:

$$CH_3COONa(aq) \longrightarrow CH_3COO^-(aq) + Na^+(aq)$$

The Na^+ ion does not react with water and is present as a spectator ion. The acetate ion (CH_3COO^-), on the other hand, is the conjugate base of the weak acid, CH_3COOH. As a weak base, the acetate ion will react with water, forming acetic acid and hydroxide ions:

$$CH_3COO^-(aq) + H_2O(l) \rightleftharpoons CH_3COOH(aq) + OH^-(aq)$$

This reaction shows that a sodium acetate solution is basic. The equilibrium constant for this reaction, which can be treated as the base ionization constant (K_b) for CH_3COO^-, is given by

$$K_b = \frac{[CH_3COOH][OH^-]}{[CH_3COO^-]} = 5.6 \times 10^{-10}$$

Calculate the pH of a 0.34 M CH_3COONa solution.

Answer *We follow the same procedure used for the ammonia solution shown earlier.*

　Step 1: *Initially we have CH_3COONa and H_2O.*

8.4

EXAMPLE

170

Step 2: *The reactions that occur in solution are*

$$CH_3COONa(aq) \longrightarrow CH_3COO^-(aq) + Na^+(aq)$$

$$CH_3COO^-(aq) + H_2O(l) \rightleftharpoons CH_3COOH(aq) + OH^-(aq)$$

$$H_2O(l) + H_2O(l) \rightleftharpoons H_3O^+(aq) + OH^-(aq)$$

Step 3: *At equilibrium, the major species in solution are* CH_3COO^-, CH_3COOH, OH^-, Na^+, *and* H_2O. *The* Na^+ *ion is a spectator ion, and we ignore the ionization of water.*

Step 4: *Let x M be the amount of the acetate ion that undergoes hydrolysis. We write*

$$CH_3COO^-(aq) + H_2O(l) \rightleftharpoons CH_3COOH(aq) + OH^-(aq)$$

Initial (M):	0.36	0	0
Change (M):	$-x$	$+x$	$+x$
Equilibrium (M)	$0.36 - x$	x	x

The base ionization constant is given by

$$K_b = \frac{[CH_3COOH][OH^-]}{[CH_3COO^-]}$$

$$5.6 \times 10^{-10} = \frac{x^2}{0.36 - x}$$

Applying the assumption $0.36 - x \approx 0.36$, *we write*

$$x^2 = (0.36)(5.6 \times 10^{-10}) = 2.0 \times 10^{-10}$$

$$x = 1.4 \times 10^{-5} M = [OH^-]$$

To check the validity of our approximation, we write

$$\frac{1.4 \times 10^{-5} M}{0.36 M} \times 100\% = 3.9 \times 10^{-3}\%$$

which is well within the acceptable limit.

Next,

$$[H_3O^+] = \frac{K_W}{[OH^-]}$$

$$= \frac{1.0 \times 10^{-14}}{1.4 \times 10^{-5}} = 7.1 \times 10^{-10} M$$

Finally,

$$pH = -log(7.1 \times 10^{-10})$$

$$= 9.1$$

As predicted, a sodium acetate solution is basic. Anytime you have a solution containing a salt in which the cation is a spectator ion and the anion is the conjugate base of a weak acid, you will solve for pH exactly as we did for sodium acetate.

A common cation that is not a spectator ion in water is the ammonium ion, NH_4^+, because it is the conjugate acid of the weak base ammonia NH_3. Let's consider what happens when ammonium chloride (NH_4Cl) is dissolved in water. The first step involves the dissociation of the salt in solution:

$$NH_4Cl(aq) \longrightarrow NH_4^+(aq) + Cl^-(aq)$$

The Cl^- ion is the conjugate base of the strong acid HCl. As a result, it does not react with water and is present as a spectator ion. The ammonium ion (NH_4^+) does react with water to a certain extent:

$$NH_4^+(aq) + H_2O(l) \rightleftharpoons NH_3(aq) + H_3O^+(aq)$$

and the acid ionization constant is given by

$$K_a = \frac{[NH_3][H_3O^+]}{[NH_4^+]}$$

Thus, an ammonium chloride solution is acidic due to the reaction of the ammonium ion with water.

A more complicated, but not commonly encountered, case is when *both* the cation and the anion of a salt react with water. Consider ammonium cyanide (NH_4CN). The cation (NH_4^+) is the conjugate acid of the weak base (NH_3), and the anion (the cyanide ion, CN^-) is the conjugate base of the weak acid (hydrocynaic acid, HCN). In solution, NH_4CN dissociates into NH_4^+ and CN^- ions

$$NH_4CN(aq) \longrightarrow NH_4^+(aq) + CN^-(aq)$$

which then react with water as follows:

$$NH_4^+(aq) + H_2O(l) \rightleftharpoons NH_3(aq) + H_3O^+(aq)$$

$$CN^-(aq) + H_2O(l) \rightleftharpoons HCN(aq) + OH^-(aq)$$

Because both H_3O^+ and OH^- ions are produced, how will the pH of the solution change? To answer this question, we need to look at the K_a value for NH_4^+ (5.6×10^{-10}) and the K_b value for CN^-, which is 2.0×10^{-5}. Because $K_b > K_a$, we conclude that the CN^- ions will hydrolyze to a greater extent than the NH_4^+ ions, so there will be more OH^- ions present than H_3O^+ ions at equilibrium. Consequently, the solution will be basic, with a pH > 7.

An important relationship exists between the K_b of a conjugate base and the K_a of the acid. In the case of the acetic acid/acetate pair, for the acid we have

$$CH_3COOH(aq) + H_2O(l) \rightleftharpoons H_3O^+(aq) + CH_3COO^-(aq)$$

and

$$K_a = \frac{[H_3O^+][CH_3COO^-]}{[CH_3COOH]}$$

And for the base we have

$$CH_3COO^-(aq) + H_2O(l) \rightleftharpoons CH_3COOH(aq) + OH^-(aq)$$

and

$$K_b = \frac{[CH_3COOH][[OH^-]}{[CH_3COO^-]}$$

Now if we multiply the two expressions together

$$K_a K_b = \frac{[H_3O^+][CH_3COO^-]}{[CH_3COOH]} \times \frac{[CH_3COOH][OH^-]}{[CH_3COO^-]}$$

$$K_a K_b = [H_3O^+][OH^-]$$
$$= K_w$$

Thus, a knowledge of K_a enables us to calculate K_b, and vice versa. Here the base ionization constant of the acetate ion is given by

$$K_b = \frac{K_w}{K_a}$$

$$= \frac{1.0 \times 10^{-14}}{1.8 \times 10^{-5}} = 5.6 \times 10^{-10}$$

Similarly, using the K_b expression for NH_3 and the K_a expression for its conjugate acid, NH_4^+, we write

$$K_a K_b = \frac{[NH_3][H_3O^+]}{[NH_4^+]} \times \frac{[NH_4^+][OH^-]}{[NH_3]}$$

$$= [H_3O^+][OH^-] = K_w$$

So for any conjugate acid-base pair, the product $K_a K_b$ is *always* equal to K_w. You can see this is the case by examining the K_a and K_b values of conjugate pairs in Tables 8.4 and 8.5.

(Check the website for problems on acid and base ionizations.)

Buffer Solutions

A **buffer solution** is a solution that resists changes in pH upon the addition of small amounts of either acid or base. Buffers are very important to chemical and biological systems. The pH in the human body varies greatly from one fluid to another. For example, the pH of blood is about 7.4, whereas the gastric juice in the stomach has a pH of about 1.5, which may decrease even further during the exam period. These pH values, which are crucial for biological functions, are maintained by buffers in most cases.

A buffer solution must contain a relatively large concentration of acid to react with any OH^- ions that are added to it and must contain a similar concentration of base to react with any added H_3O^+ ions. Furthermore, the acid and the base components of the buffer must not consume each other in a neutralization reaction. These requirements are satisfied by an acid-base conjugate pair; for example, a weak acid and its conjugate base (supplied by a salt) or a weak base and its conjugate acid (supplied by a salt).

A simple buffer solution can be prepared by adding comparable amounts of acetic acid (CH_3COOH) and its salt sodium acetate (CH_3COONa) to water. The conjugate base is the acetate ion, CH_3COO^-, formed by the dissociation of sodium acetate:

$$CH_3COONa(aq) \longrightarrow CH_3COO^-(aq) + Na^+(aq)$$

There are two things to note about this buffer system. First, the acid and the base do not neutralize each other because such a "neutralization" produces the same substances:

$$CH_3COOH(aq) + CH_3COO^-(aq) \rightleftharpoons CH_3COO^-(aq) + CH_3COOH(aq)$$

so no net change occurs. Second, the ionization of the acid

$$CH_3COOH(aq) + H_2O(l) \rightleftharpoons H_3O^+(aq) + CH_3COO^-(aq)$$

and the hydrolysis of the acetate ion

$$CH_3COO^-(aq) + H_2O(l) \rightleftharpoons CH_3COOH(aq) + OH^-(aq)$$

can generally be ignored. According to Le Chatelier's principle, the presence of the CH_3COO^- ions (from CH_3COONa) would shift the ionization reaction from right to left, and the presence of CH_3COOH (from the original acid) would shift the hydrolysis reaction also from right to left. Consequently, we can treat the equilibriun concentrations of acetic acid and acetate ions as the initial concentrations. This approach greatly simplifies calculations involving buffers.

How does such a buffer solution work? Suppose we add some NaOH to it. As you know, the presence of NaOH in solution produces OH^- ions. These ions will be neutralized by the acid present, that is

$$CH_3COOH(aq) + OH^-(aq) \longrightarrow CH_3COO^-(aq) + H_2O(l)$$

Such a reaction will decrease the concentration of CH_3COOH and increase the concentration of CH_3COO^-. But as we will demonstrate shortly, it will not significantly affect the pH of the solution as long as the amount of NaOH added is not too great. Similarly, if we add a small amount of HCl to the buffer, the excess H_3O^+ ions provided by the hydrochloric acid will be neutralized by the acetate ions:

$$CH_3COO^-(aq) + H_3O^+(aq) \longrightarrow CH_3COOH(aq) + H_2O(l)$$

In this case, there will be a decrease in the CH_3COO^- concentration and an increase in the CH_3COOH concentration. The pH of the solution, however, will remain largely unchanged.

EXAMPLE 8.5

(a) Calculate the pH of a buffer system containing 1.0 M CH_3COOH and 1.0 M CH_3COONa. (b) Calculate the pH after the addition of 0.10 mole of HCl to 1 liter of the buffer solution.

Answer *(a) To calculate the pH of the buffer solution, note that we can treat the equilibrium concentrations of the acid and the base as the initial concentrations. Therefore, at equilibrium,*

$$[CH_3COOH] = 1.0\ M \quad and \quad [CH_3COO^-] = 1.0\ M$$

Next, we write the acid ionization constant for acetic acid

$$K_a = \frac{[H_3O^+][CH_3COO^-]}{[CH_3COOH]}$$

and rearrange to get $[H_3O^+]$

$$[H_3O^+] = \frac{K_a[CH_3COOH]}{[CH_3COO^-]}$$

$$= \frac{(1.8 \times 10^{-5})(1.0)}{(1.0)}$$

$$= 1.8 \times 10^{-5}\ M$$

—Continued next page

Continued—

Therefore,

$$pH = -log\ (1.8 \times 10^{-5})$$

$$= 4.74$$

Note that because K_a is an equilibrium constant, its value is unchanged whether CH_3COOH and CH_3COO^- come from the same source (the acid) or two different sources (the acid and sodium acetate), as in a buffer solution.

(b) From the initial concentrations, we see that in 1 liter of the solution, there is 1.0 mole of CH_3COOH and 1.0 mole of CH_3COO^-. The addition of HCl produces 0.10 mole of H_3O^+ ions:

$$HCl(aq) + H_2O(l) \longrightarrow H_3O^+(aq) + Cl^-(aq)$$
$$0.10\ mol \qquad\qquad 0.10\ mol \quad 0.10\ mol$$

The Cl^- ion is a spectator ion. The H_3O^+ ions produced are neutralized by the conjugate base acetate ion as follows:

$$CH_3COO^-(aq) + H_3O^+(aq) \longrightarrow CH_3COOH(aq) + H_2O(l)$$
$$0.10\ mol \qquad\quad 0.10\ mol \qquad\qquad 0.10\ mol$$

At equilibrium, the adjusted moles of the acid and base are

$$CH_3COOH: 1.0\ mol + 0.10\ mol = 1.1\ mol$$

$$CH_3COO^-: 1.0\ mol - 0.10\ mol = 0.90\ mol$$

Using the expression in (a), we calculate $[H_3O^+]$ by writing

$$[H_3O^+] = \frac{K_a[CH_3COOH]}{[CH_3COO^-]}$$

$$= \frac{(1.8 \times 10^{-5})(1.1)}{0.90} = 2.2 \times 10^{-5}\ M$$

Note that because both CH_3COOH and CH_3COO^- are in the same solution, their concentration (moles per liter) ratio is the same as the ratio of number of moles. The pH of the solution is

$$pH = -log\ (2.2 \times 10^{-5})$$

$$= 4.66$$

Thus, the decrease in pH unit is 4.74 − 4.66 = 0.08.

To see how effective this buffer solution is in maintaining a fairly constant pH, we can compare the changes in $[H_3O^+]$ as follows:

$$\text{before addition of HCl: } [H_3O^+] = 1.8 \times 10^{-5}\ M$$

$$\text{after addition of HCl: } [H_3O^+] = 2.2 \times 10^{-5}\ M$$

Thus, the H_3O^+ ion concentration increases by a factor of

$$\frac{2.2 \times 10^{-5}\ M}{1.8 \times 10^{-5}\ M} = 1.2$$

What would be the change in pH and $[H_3O^+]$ if 0.10 mole of HCl were added to 1 liter of pure water? In this case, we have

before addition of HCl: $[H_3O^+] = 1.0 \times 10^{-7}\ M$ pH = 7.00

after addition of HCl: $[H_3O^+] = 0.10\ M$ pH = 1.00

so there will be a change in pH of six units and the $[H_3O^+]$ would increase by a factor of

$$\frac{0.10\ M}{1.0 \times 10^{-7}M} = 1.0 \times 10^6$$

amounting to a million-fold increase!

Here are some facts about buffer solutions you should understand:

1. A buffer solution must contain a weak acid and its conjugate base. Strong acids would not function in a buffer system. Consider a solution of HCl and NaCl (containing the conjugate base Cl^-). The H_3O^+ ions derived from HCl would neutralize an added base,

$$H_3O^+(aq) + OH^-(aq) \longrightarrow 2H_2O(l)$$

but the conjugate base cannot neutralize an added acid,

$$Cl^-(aq) + H_3O^+(aq) \longleftarrow HCl(aq) + H_2O(l)$$

because HCl is a strong acid and therefore the reaction would not proceed from left to right.
2. An effective buffer must have comparable amounts of the acid and base components, and the concentrations must be sufficiently high (0.1 M or higher).

Preparing a Buffer Solution with a Specific pH

Now the question arises. How does one go about preparing a buffer at a specific pH? In other words, of the many weak acids and their conjugate bases, which combination should one choose? Consider the ionization of the weak acid HB:

$$HB(aq) + H_2O(l) \rightleftharpoons H_3O^+(aq) + B^-(aq)$$

The ionization constant is given by

$$K_a = \frac{[H_3O^+][B^-]}{[HB]}$$

Rearranging, we get

$$[H_3O^+] = K_a \frac{[HB]}{[B^-]}$$

We now take the negative of the logarithm of the equation

$$-\log [H_3O^+] = -\log K_a - \log \frac{[HB]}{[B^-]}$$

$$pH = pK_a + \log \frac{[B^-]}{[HB]} \qquad (8.1)$$

where

$$pK_a = -\log K_a$$

176

Equation (8.1) is called the *Henderson-Hasselbalch equation*. It holds whether we just have the pure acid or a mixture of the acid and the conjugate base (from a salt like NaA) in solution. In general, we can write

$$pH = pK_a + \log \frac{[\text{conjugate base}]}{[\text{acid}]} \qquad (8.2)$$

Recall that for a buffer to function effectively, the concentration of the acid and its conjugate base must be roughly the same, that is, [conjugate base] ≈ [acid]. Because log 1 = 0, this means that

$$\log \frac{[\text{conjugate base}]}{[\text{acid}]} \approx 0$$

so that

$$pH \approx pK_a$$

It follows, therefore, that whenever we wish to prepare a buffer solution of a desired pH, we must choose an acid whose pK_a value closely matches the pH value.

8.6

EXAMPLE

A student is asked to prepare a buffer solution at pH = 8.60, using one of the following weak acids: HA ($K_a = 2.7 \times 10^{-3}$), HB ($K_a = 4.4 \times 10^{-6}$), HC ($K_a = 2.6 \times 10^{-9}$). Which acid should she choose? Why?

Answer *We want an acid whose pK_a value is roughly equal to the desired pH. Therefore, the first step is to convert the K_a values to pK_a:*

$$HA \quad pK_a = -\log 2.7 \times 10^{-3} = 2.57$$
$$HB \quad pK_a = -\log 4.4 \times 10^{-6} = 5.36$$
$$HC \quad pK_a = -\log 2.6 \times 10^{-9} = 8.59$$

It is clear, then, the buffer system should consist of HC and its salt (NaC) because the pK_a and pH values are nearly the same.

As an exercise, let's see what would happen if we had chosen HA and NaA to prepare the buffer solution at pH = 8.60. From the Henderson-Hasselbalch equation, we write

$$8.60 = 2.57 + \log \frac{[\text{conjugate base}]}{[\text{acid}]}$$

$$\log \frac{[\text{conjugate base}]}{[\text{acid}]} = 6.03$$

$$\frac{[\text{conjugate base}]}{[\text{acid}]} = \text{antilog } 6.03 = 10^{-6.03} = 9.3 \times 10^{-7}$$

or

$$[\text{conjugate base}] = 9.3 \times 10^{-7}[\text{acid}]$$

Thus the concentration of the acid is about a million times that of the conjugate base. Such a "buffer solution" would be totally ineffective toward an added acid because the concentration of the conjugate base would be too small.

(Check the website for more problems on buffer solutions.)

Test your understanding of the material in this chapter

IMPORTANT TERMS

Explain the following terms in your own words:

acid ionization constant (K_a), p. 165
base ionization constant (K_b), p. 167
buffer solution, p. 172
chemical equilibrium, p. 146
conjugate acid, p. 160
conjugate base, p. 160
equilibrium constant (K), p. 151
law of mass action, pp. 152, 158
Le Chatelier's principle
pH, p. 161
reaction quotient, p. 155
salt hydrolysis, p. 169

UNDERSTANDING CHEMISTRY

Summarizing Problem

(a) Calculate the pH of a 0.10 *M* nitrous acid (HNO_2) solution. (b) What is the pH after the addition of 0.10 mole of sodium nitrite ($NaNO_2$) to 1 liter of the acid solution? (c) What is the pH of the solution in (b) after the addition of 0.050 mole of sodium hydroxide (NaOH)? Assume no change in volume in both (b) and (c). The ionization constant of HNO_2 is 4.5×10^{-4}.

Answers: (a) We follow Steps 1–4 to solve for the pH of the solution.

Step 1: Initially we have HNO_2 and H_2O present.

Step 2: The reactions that occur in solution are

$$HNO_2(aq) + H_2O(l) \rightleftharpoons H_3O^+(aq) + NO_2^-(aq)$$

$$H_2O(l) + H_2O(l) \rightleftharpoons H_3O^+(aq) + OH^-(aq)$$

Step 3: At equilibrium, the major species in solution are HNO_2, H_3O^+, NO_2^-, and H_2O. Note that even though HNO_2 is a weak acid, it is the main source of H_3O^+ ions, and we can ignore the ionization of water.

Step 4: We now solve for $[H_3O^+]$ using the ICE method. Let x *M* be the amount of HNO_2 that undergoes ionization so that

	$HNO_2(aq)$ + $H_2O(l)$ $\rightleftharpoons$	$H_3O^+(aq)$ +	$NO_2^-(aq)$
Initial (*M*):	0.10	0	0
Change (*M*):	$-x$	$+x$	$+x$
Equilibrium (*M*):	$0.10 - x$	x	x

The acid ionization constant is given by

$$K_a = \frac{[H_3O^+][NO_2^-]}{[HNO_2]}$$

$$4.5 \times 10^{-4} = \frac{x^2}{0.10 - x}$$

Remember that in solving problems like this, we always try the shortcut first. Because HNO_2 ionizes only to a slight extent, x must be small compared to 0.10. Therefore, we can make the approximation

$$0.10 - x \approx 0.10$$

Now the acid ionization constant expression becomes

$$4.5 \times 10^{-4} = \frac{x^2}{0.10}$$

Rearranging, we get

$$x^2 = (0.10)(4.5 \times 10^{-4})$$

$$x = \sqrt{4.5 \times 10^{-5}} = 6.7 \times 10^{-3} \, M$$

Checking the validity of our approximation, we write

$$\frac{6.7 \times 10^{-3} \, M}{0.10 \, M} \times 100\% = 6.7\%$$

As you can see, the answer is greater than 5%, so the approximation does not hold. Next, we write the quadratic equation as

$$x^2 + 4.5 \times 10^{-4}x - 4.5 \times 10^{-5} = 0$$

Solving, we get

$$x = -6.9 \times 10^{-3} \, M \quad \text{or} \quad x = 6.5 \times 10^{-3} \, M$$

The first answer is physically impossible (why?), so the pH of the solution is given by

$$pH = -\log 6.5 \times 10^{-3}$$

$$= 2.19$$

(b) According to Le Chatelier's principle, the increase in the nitrite ion concentration will suppress the ionization of HNO_2. Therefore, we expect an increase in the pH of the solution. In fact, we have a buffer system at this point. Rearranging the ionization expression for HNO_2 shown previously, we write

$$[H_3O^+] = \frac{K_a[HNO_2]}{[NO_2^-]}$$

If we treat the equilibrium concentrations of HNO_2 and NO_2^- the same as their initial concentrations, then

$$[H_3O^+] = \frac{(4.5 \times 10^{-4})(0.10)}{0.10}$$

$$= 4.5 \times 10^{-4} \, M$$

Finally, the pH is given by

$$pH = -\log 4.5 \times 10^{-4}$$

$$= 3.35$$

(You may remember that when the concentration of the acid is equal to the concentration of its conjugate base, the pH of the solution is numerically equal to the pK_a of the acid.)

(c) NaOH is a strong electrolyte and dissociates completely in solution. Therefore, the initial amount of the hydroxide ions in solution is 0.050 mole, and the neutralization reaction that occurs is

$$\begin{array}{ccccc} HNO_2 & + & OH^- & \longrightarrow H_2O \,\, + & NO_2^- \\ 0.050 \text{ mol} & & 0.050 \text{ mol} & & 0.050 \text{ mol} \end{array}$$

At equilibrium, the concentrations of the acid and conjugate base are $[HNO_2] = 0.050$ M and $[NO_2^-] = 0.15 \, M$. Using the Henderson-Hasselbalch equation, we calculate the pH of the solution

$$pH = 3.35 + \log \frac{0.15}{0.050}$$

$$= 3.83$$

As we would expect, the pH of the solution increases because a base has been added to it, although the change is relatively small considering half of the acid has been consumed in the neutralization reaction.

Glossary

A

acid. A substance that yields hydrogen ions (H^+) when dissolved in water. (p. 22)

acid-base reaction. Reaction between an acid and a base. Also called acid-base neutralization if the reaction leads to the formation of a salt and water. (p. 39)

acid-base neutralization reaction. A reaction between an acid and a base to produce a salt and water. (p. 39)

acid ionization constant. The equilibrium constant for acid ionization. (p. 165)

actual yield. The amount of product actually obtained in a reaction. (p. 54)

alkali metals. The Group 1A elements (Li, Na, K, Rb, Cs, and Fr). (p. 15)

alkaline earth metals. The Group 2A elements (Be, Mg, Ca, Sr, Ba, and Ra). (p. 15)

allotropes. Two or more forms of the same element that differ significantly in chemical and physical properties. (p. 73)

angular momentum quantum number. Designates the number of atomic orbitals associated with a given principal quantum number *n*. (p. 89)

anion. An ion with a net negative charge. (p. 17)

atom. The basic unit of an element that can enter into chemical combination. (p. 9)

atomic mass. The mass of an atom in atomic mass units. (p. 47)

atomic mass unit (amu). A mass exactly equal to one-twelfth the mass of one carbon-12 atom. (p. 47)

atomic number. The number of protons in the nucleus of an atom. (p. 11)

atomic radius. One-half the distance between the two nuclei in two adjacent atoms in a metal or one-half the distance between the nuclei of two atoms of the same element in a diatomic molecule. (p. 98)

Avogadro, Lorenzo. Italian scientist whose law formed the basics for determining atomic masses of the elements. (p. 48)

Avogadro's number. 6.022×10^{23}; the number of particles in a mole. (p. 48)

B

Balmer, Johann. Swiss mathematician who formulated an equation for analyzing the hydrogen emission spectra. (p. 84)

base. A substance that yields hydroxide ions (OH^-) when dissolved in water. (p. 24)

base ionization constant. The equilibrium constant for the ionization of a base. (p. 167)

Bohr, Niels. Danish physicist who proposed a model for the hydrogen atom to explain its emission spectra. (p. 85)

Brønsted, Johannes. Danish chemist who defined acids as proton donors and bases as proton acceptors. (p. 35)

Brønsted acid. A substance capable of donating a proton in a reaction. (p. 35)

Brønsted base. A substance capable of accepting a proton in a reaction. (p. 35)

buffer solution. A solution of (a) a weak acid or base and (b) its salt; both components must be present. A buffer solution has the ability to resist changes in pH when small amounts of either acid or base are added to it. (p. 172)

C

calorimeter. A device for measuring the heat change of a chemical reaction or a physical process. (p. 67)

calorimetry. The measurement of heat changes. (p. 67)

cation. An ion with a net positive charge. (p. 17)

chemical equation. An equation that uses chemical symbols to show what happens during a chemical reaction. (p. 29)

chemical equilibrium. A state in which the rates of the forward and reverse reactions are equal and no net changes can be observed. (pp. 37, 146)

chemical formula. An expression showing the chemical composition of a compound in terms of the symbols for the atoms of the elements involved. (p. 16)

chemical nomenclature. Naming of chemical compounds. (p. 19)

chemical property. Any property of a substance that cannot be studied without converting the substance into some other substance. (p. 25)

chemical reaction. Chemical change. (p. 9)

compound. A substance composed of atoms of two or more elements chemically united in fixed proportions. (p. 9)

concentration. The amount of solute present in a given quantity of solution. (p. 56)

conjugate acid. The acid formed when a base accepts a proton. (p. 160)

conjugate base. The base formed when an acid loses a proton. (p. 160)

covalent bond. A bond in which two electrons are shared by two atoms. (p. 63)

D

Dalton, John. English scientist and school teacher who first formulated a precise definition for atoms. (p. 9)

de Broglie, Louis. French physicist who proposed that matter and radiation have both wave and particle properties. (p. 86)

Debye, Peter. American chemist and physicist of Dutch origin who made significant contributions to the study of molecular structure. (p. 135)

Democritus. Greek philosopher who first gave a description for atoms. (p. 9)

diamagnetic. A substance containing only paired electrons.(p. 97)

diatomic molecule A molecule that consists of two atoms. (p. 16)

dipole-dipole forces. Forces that act between polar molecules. (p. 139)

dipole moment. The product of charge and the distance between the charges in a molecule. (p. 135)

diprotic acid. An acid in which each unit of the acid yields two hydrogen ions. (p. 24)

dispersion forces. The attractive forces that arise as a result of temporary dipoles induced in the atoms or molecules. (p. 142)

E

Einstein, Albert. German-born American physicist who introduced the particle concept of light (photons) to explain the photoelectric effect. (p. 82)

electrolyte. A substance that, when dissolved in water, results in a solution that can conduct electricity. (p. 25)

electron. A subatomic particle that has a very low mass and carries a single negative electric charge. (p. 10)

electron affinity. A measure of the affinity of a gaseous atom for an electron. (p. 98)

electron configuration. The distribution of electrons among the various orbitals in an atom or molecule. (p. 94)

electronegativity. The ability of an atom to attract electrons toward itself in a chemical bond. (p. 106)

electron spin quantum number. A number $(+\frac{1}{2})$ or $(-\frac{1}{2})$ that describes the spinning motion of the electron. (p. 90)

element. A substance that cannot be separated into simpler substances by chemical means. (p. 9)

endothermic processes. Processes that absorb heat from the surroundings. (p. 66)

energy. The capacity to do work or to produce change. (p. 61)

enthalpy. A thermodynamic quantity used to describe heat changes taking place at constant pressure. (p. 70)

equilibrium constant (K). A number equal to the ratio of the equilibrium concentrations of products to the equilibrium concentrations of reactants, each raised to the power of its stoichiometric coefficient. (p. 151)

excess reagent. A reactant present in a quantitiy greater than necessary to react with the amount of the limiting reagent present. (p. 53)

excited level. An energy level that has higher energy than the ground level. (p. 85)

excited state. A state that has higher energy than the ground state. (p. 85)

exothermic processes. Processes that give off heat to the surroundings. (p. 66)

expanded octet. The number of valence electrons surrounding a central atom exceeds eight. (p. 125)

F

family. The elements in a vertical column of the periodic table. (p. 13)

formal charge. The difference between the valence electrons in an isolated atom and the number of electrons assigned to that atom in a Lewis structure. (p. 122)

formula unit mass. The mass in amu of one unit of an ionic compound. (p. 47)

G

Geiger, Hans. German physicist who worked with Rutherford on atomic structure. He also invented a device for measuring radiation that is now commonly called the Geiger counter.(p. 11)

ground level. The lowest-energy level of a system. (p. 85)

ground state. The lowest-energy state of a system. (p. 85)

group. The elements in a vertical column of the periodic table. (p. 13)

Guldberg, Cato. Norwegian chemist who formulated the law of mass action with Peter Waage. (p. 152)

H

half-reaction. Reaction showing either the oxidation or reduction of the redox process. (p. 40)

halogens. The nonmetallic elements in Group 7A (F, Cl, Br, I, and At). (p. 15)

heat. Transfer of energy between two bodies that are at different temperatures. (p. 63)

Hess, Germain. Swiss chemist who formulated the law of heat summation, which is now known as Hess's law. (p. 72)

Hess's law. When reactants are converted to products, the change in enthalpy is the same whether the reaction takes place in one step or in a series of steps. (p. 72)

Hund, Frederick. German physicist who formulated the rule for predicting the number of parallel electron spins in an atom. (p. 97)

Hund's rule. The most stable arrangement of electrons in atomic subshells is the one with the greatest number of parallel spins. (p. 97)

hydrogen bonding. A special type of dipole-dipole interaction between the hydrogen atom bonded to an atom of a very electronegative element (F, N, O) and another atom of one of the three electronegative elements. (p. 139)

hydronium ion. H_3O^+. (p. 36)

I

ideal gas equation. An equation expressing the relationships among pressure, volume, temperature, and amount of gas ($PV = nRT$, where R is the gas constant). (p. 57)

induced dipole. The separation of positive and negative charges in an atom (or a nonpolar molecule) caused by the proximity of an ion or a polar molecule. (p. 141)

intermolecular forces (IMFs). Attractive forces that exist among molecules. (pp. 64, 138)

ion. A charged particle formed when a neutral atom or group of atoms gains or loses one or more electrons. (p. 17)

ion-dipole forces. Forces that operate between an ion and a dipole. (p. 138)

ionic bond. The electrostatic force that holds ions together in an ionic compound. (p. 64)

ionic compound. Any neutral compound containing cations and anions. (p. 17)

ionic equation. An equation that shows dissolved ionic compounds in terms of their free ions. (p. 34)

ionic radius. The radius of a cation or an anion. (p. 99)

ionization energy. The minimum energy required to remove an electron from an isolated atom (or an ion) in its ground state. (p. 98)

isoelectronic. Having the same number of electrons. (p. 96)

isotope. Atoms having the same atomic number but different mass numbers. (p. 11)

K

Kelvin, Lord William Thomson. Scottish physicist who worked on atomic structure and introduced the absolute temperature scale. (p. 84)

L

law of conservation of energy. The total quantity of energy in the universe is constant. (p. 65)

law of conservation of mass. Matter can be neither created nor destroyed. (p. 29)

law of mass action. For a reversible reaction at equilibrium and constant temperature, a certain ratio of product and reaction concentrations has a constant value, called the equilibrium constant. (p. 152)

Le Chatelier, Henri. French chemist who formulated the principle for predicting how a system at equilibrium would respond to a stress applied to the system. (p. 157)

Le Chatelier's principle. If an external stress is applied to a system at equilibrium, the system will adjust itself in such a way as to partially offset the stress as it reaches a new equilibrium position. (p. 158)

Lewis, Gilbert. American chemist who formulated the covalent bond and the octet rule. (p. 108)

Lewis structure. A representation of covalent bonding using Lewis symbols. Shared electron pairs are shown either as lines or as pairs of dots between two atoms, and lone pairs are shown as pairs of dots on individual atoms. (p. 110)

limiting reagent. The reactant used up first in a reaction. (p. 53)

lone pair. Valence electrons that are not involved in covalent bond formation. (p. 110)

M

magnetic quantum number. A quantum number that describes the orientation of orbitals in space. (p. 90)

Marsden, Ernest. English physicist who, as an undergraduate, worked with Rutherford on atomic structure. (p. 11)

mass number. The total number of neutrons and protons in the nucleus of an atom. (p. 11)

Mendeleev, Dmitri. Russian chemist who devised the periodic classification of elements. (p. 13)

metal. An element that is a good conductor of heat and electricity and has the tendency to form positive ions in ionic compounds. (p. 14)

metalloid. An element with properties intermediate between those of metals and nonmetals. (p. 14)

Millikan, Robert. American physicist who determined the charge of an electron. (p. 83)

miscible. Two liquids that are completely soluble in each other in all proportions. (p. 143)

molarity. The number of moles of solute in 1 liter of solution. (p. 56)

molar mass. The mass (in grams or kilograms) of 1 mole of atoms, molecules, or other particles. (p. 48)

mole. The amount of substance that contains as many elementary entities (atoms, molecules, or other particles) as there are atoms in exactly 12 grams (or 0.012 kilograms) of the carbon-12 isotope. (p. 48)

molecular compound. A molecule composed of atoms of two or more elements. (p. 20)

molecular mass. The sum of the atomic masses (in amu) present in a given molecule. (p. 47)

molecular orbital theory. A theory of chemical bond formation. (p. 111)

molecule. An aggregate of at least two atoms in a definite arrangement held together by chemical forces. (p. 15)

monatomic ion. An ion that contains only one atom. (p. 18)

monoprotic acid. An acid in which each unit of the acid yields one hydrogen ion. (p. 24)

N

net ionic equation. An equation that includes only the ionic species that actually take part in the reaction. (p. 34)

neutron. A subatomic particle that bears no net electric charge. Its mass is slightly greater than a proton's. (p. 10)

noble gases. Nonmetallic elements in Group 8A (He, Ne, Ar, Kr, Xe, and Rn). (p. 15)

nonelectrolyte. A substance that, when dissolved in water, gives a solution that is not electrically conducting. (p. 25)

nonmetal. An element that is usually a poor conductor of heat and electricity. (p. 14)

nucleus. The central core of an atom. (p. 9)

O

octet rule. An atom other than hydrogen tends to form bonds until it is surrounded by eight valence electrons. (p. 116)

orbital. Quantum mechanical description of region where an electron in an atom can be found. (p. 87)

orbital diagram. A diagram showing the placement of electrons in various atomic orbitals. (p. 97)

oxidation. A loss of electrons or an increase in oxidation number. (p. 40)

oxidation number. The number of charges an atom would have in a molecule if electrons were transferred completely in the direction indicated by the difference in electronegativity. (p. 42)

oxidation-reduction reaction. A reaction involving gain and loss of electrons or decrease and increase in oxidation number of the elements. (p. 40)

oxidizing agent. A substance that can accept electrons from another substance or increase the oxidation numbers of another substance. (p. 41)

oxoacid. An acid containing hydrogen, oxygen, and another element (the central element). (p. 23)

oxoanion. An anion derived from an oxoacid. (p. 23)

P

paramagnetic. A paramagnetic substance contains one or more unpaired electrons. (p. 97)

Pauli, Wolfgang Austrian physicist who formulated the exclusion principle that restricts the quantum numbers an electron can possess in an atom. (p. 88)

Pauli exclusion principle. No two electrons in an atom can have the same four quantum numbers. (p. 88)

Pauling, Linus. American chemist who introduced the concept of electronegativity and worked on the nature of chemical bonding. (p. 106)

percent yield. The ratio of the actual yield of a reaction to the theoretical yield, multiplied by 100%. (p. 54)

period. A horizontal row of the periodic table. (p. 13)

periodic group. A vertical column of the periodic table. (p. 12)

periodic table. A tabular arrangement of the elements by similarities in properties and by increasing atomic number. (p. 12)

pH. The negative logarithm of the hydrogen ion concentration in an aqueous solution. (p. 161)

photon. A particle of light. (p. 83)

physical property. Any property of a substance that can be observed without transforming the substance into some other substance. (p. 25)

Planck, Max. German physicist whose work on black-body radiation (quantization of energy) marked the beginning of quantum theory. (p. 82)

polar molecule. A molecule that possesses a dipole moment. (pp. 26, 105)

polyatomic ion. An ion that contains more than one atom. (p. 18)

polyatomic molecule. A molecule that consists of more than two atoms. (p. 16)

precipitate. An insoluble solid that separates from a solution. (p. 33)

precipitation reaction. A reaction characterized by the formation of a precipitate. (p. 33)

principal quantum number. A quantum number that determines the size of an orbital and its energy. (p. 89)

product. The substance formed as a result of a chemical reaction. (p. 30)

proton. A subatomic particle having a single positive electric charge. The mass of a proton is about 1840 times that of an electron. (p. 9)

Q

quanta. The smallest quantity of energy that can be emitted or absorbed in the form of electromagnetic radiation. (p. 82)

quantum mechanics. A scientific discipline that deals with the behavior of atoms and molecules. (p. 87)

quantum number. Numbers that describe the distribution of electrons in atoms. (p. 88)

quantum theory. A theory that deals with the behavior of light and atomic behavior. (p. 82)

R

reactant. The starting substances in a chemical reaction. (p. 30)

reaction quotient. A number equal to the ratio of product concentrations to reactant concentrations, each raised to the power of its stoichiometric coefficient at some point other than equilibrium. (p. 155)

redox reaction An oxidation-reduction reaction. (p. 40)

reducing agent. A substance that can donate electrons to another substance or decrease the oxidation numbers in another substance. (p. 41)

reduction. A gain of electrons or a decrease in oxidation number. (p. 40)

representative elements. Elements in Groups 1A through 7A, all of which have at least one incompletely filled s or p subshell of the highest principal quantum number. (p. 21)

resonance. The use of two or more Lewis structures to represent a particular molecule. (p. 119)

resonance structure. One of two or more alternative Lewis structures for a molecule that cannot be described fully with a single Lewis structure. (p. 119)

Richter, Jeremias. German chemist who coined the term stoichiometry for the study of mass relationships of chemical reactions. (p. 46)

Rutherford, Ernest New Zealand physicist whose experiments show that an atom is comprised of a dense nucleus containing protons and neutrons and the rest of the atom is mostly empty space. (p. 11)

S

salt. An ionic compound made up of a cation other than H^+ and an anion other than OH^- or O^{2-}. (p. 39)

salt hydrolysis. The reaction of the cation or anion, or both, of a salt with water. (p. 169)

Schrödinger, Erwin. Austrian physicist who formulated an equation for studying the properties of electrons. (p. 87)

solubility. The maximum amount of solute that can be dissolved in a given quantity of solvent at a specific temperature. (p. 25)

solute. The substance present in the smaller amount in a solution. (p. 25)

solution. A homogeneous mixture of two or more substances. (p. 25)

solvent. The substance present in the larger amount in a solution. (p. 25)

Sorensen, Soren. Danish biochemist who introduced the pH scale. (p. 161)

specific heat.   The amount of heat energy required to raise the temperature of 1 gram of the substance by 1 degree Celsius. (p. 68)

standard enthalpy of formation.   The heat change that results when 1 mole of a compound is formed from its elements in their standard states. (p. 73)

standard state.   The condition of 1 atm of pressure. (p. 71)

stoichiometry.   The mass relationships among reactants and products in chemical reactions. (p. 46)

strong acid.   An acid that is a strong electrolyte. (p. 37)

strong base.   A base that is a strong electrolyte. (p. 38)

T

theoretical yield.   The amount of product predicted by the balanced equation when all of the limiting reagent has reacted. (p. 54)

thermochemistry.   The study of heat changes in chemical reactions. (p. 65)

Thomson, George.   British physicist whose experiments demonstrated the wave properties of electrons. (p. 87)

Thomson, Joseph John.   British physicist who measured the charge-to-mass ratio of the electron. (p. 84)

transition metals.   Elements in Group 1B and 3B–8B. (p. 22)

triprotic acid.   An acid in which each unit of the acid yields three hydrogen ions. (p. 24)

V

valence electrons.   Electrons in the outer-shell of an atom that take part in chemical bonding. (p. 100)

valence shell.   The outermost shell of an atom. (p. 98)

valence-shell electron-pair repulsion (VSEPR).   A model that accounts for the geometrical arrangements of shared and unshared electron pairs around a central atom in terms of the repulsive forces between electron pairs. (p. 129)

W

Waage, Peter.   Norwegian chemist who formulated the law of mass action with Cato Guldberg. (p. 152)

weak acid.   An acid that is a weak electrolyte. (pp. 37, 165)

weak base.   A base that is a weak electrolyte. (pp. 37, 165)

Y

Young, Thomas.   English scientist and physician who demonstrated the wave properties of light. (p. 80)

Index